Math 6–12 Tutor's Practice Resource

PRACTICE ONLY

DAMENE W WOLDEAB

Order this book online at www.trafford.com
or email orders@trafford.com

Most Trafford titles are also available at major online book retailers.

Printed in the United States of America.

ISBN: 978-1-4907-2425-6 (sc)
ISBN: 978-1-4907-2424-9 (e)

Trafford rev. 01/24/2014

Trafford PUBLISHING® www.trafford.com
North America & international
toll-free: 1 888 232 4444 (USA & Canada)
fax: 812 355 4082

Lesson	Title	Page
	Order of Operations	1
	Prime Factorization	2
	Greatest Common Factor	3
	Least Common Multiple	4
	Compare Fractions	5
	Change Repeating Fractions to Decimals	6
	Simplifying Fractions to Lowest term	7
	Reciprocals and Mixed Fractions	8
	Adding Fractions	9
	Subtracting Fractions	10
	Multiplying Fractions	11
	Dividing Fractions	12
	Writing Fractions to Decimals	13
	Writing Decimals to Fractions	14
	Ordering Numbers from Least to Greatest	15
	Writing Fractions and Decimals as Percent	16
	Solving Percent Equations	17
	Adding Decimals	18
	Subtracting Decimals	19
	Multiplying Decimals	20
	Dividing Decimals	21
	Factoring Expressions by using difference of two squares 2	44
	Factoring Quadratic Expressions 3	45
	Factoring Quadratic Expressions 4	46
	Solving Quadratic Equations 5	47
	Finding the Discriminant of a Quadratic Equation	48
	Finding the Product of Quadratic Factors	49

Lesson	Title	Page
	Finding C in Quadratic equation to make it perfect square.	50
	Solving by Completing the Square	51
	Finding the Vertex, Axis of Symmetry, Maximum or Minimum Value, and X and Y Intercepts	52
	Making Tables, Sketching Graphs, and Writing the Domain and Range for a function	53
	Finding Zeroes, if Even, Odd, or Neither of polynomial functions	54
	Calculating Average Rate of Change	55
	Finding Terms in a Sequence	56
	Writing a Rule for a Sequence	57
	Finding the Sum of a Series	58
	Writing Rules for the nth Term of an Arithmetic Sequence	59
	Writing Rules for the nth Term of a Geometric Sequence	60
	Basic Operations of Matrices 1	61
	Conics: Equation of circles 2	62
	Conics: Equations of circles 3	63
	Conics: Equations of circles 4	64
	Conics: Parabola	65
	Conics: Parabola	66
	Conics: Hyperbola	67
	Perimeter and circumference of shapes	68
	Area of shapes	69
	Area of Shapes	70
	Pythagorean theorem	71
	Naming of Angles	72
	Naming Supplementary Angles	73
	Solving Complementary Angles	74
	Solving Complementary Angles	75

Lesson	Title	Page
	Triangle Inequality Angles	76
	Triangle Inequality Sides	77
	Triangle Interior Anglest	78
	Triangle Exterior Angles	79
	Parallel Lines	80
	Parallelograms	81
	Trapezium (Trapezoid)	82
	Basic Trigonometric Ratio	83
	Reading Inverse of Angles	84
	Finding unknown sides using Trigonometric Ratio	85
	Finding unknown Angles using Trigonometric Ratio	86
	Converting Angles to Radian Measures	87
	Converting Radians to Angle Measures	88
	Circles: Finding Arcs and Central Angles, Tangent and Chords	89
	Circles: Finding Angles and lengths in Tangent-Tangent relation.	90
	Circles: Finding Angles and Lengths in Secant-Secant relation	91
	Circles: Finding Angles and Lengths in Tangent-Secant relation	92
	Circles: Finding Angles and Lengths in Chord-Chord relation	93
	Circles: Inscribed Quadrilateral	94
	Circles: Tangent and radius relation	95

Name_____ **Date** _____

Use order of operation and simplify.

1. $3(5 - 6 \times 2) \div 7 = $ _____

2. $(6 - 2 \times 1) - 5(5 - 2)^2 + 4 = $ _____

3. $5(7 - 4 \times 2) - 4 = $ _____

4. $3(5 - 9)^2 - 3^2 = $ _____

5. $-3(2 - 5)^2 = $ _____

6. $6 \div 2 + 4 - 3 = $ _____

7. $2(7 - 3)^3 - 5 \times 2 = $ _____

8. $7 - 3 + 4 \div 2 = $ _____

Name_____ **Date** _____

Find the prime factorization of the numbers or terms given below

1. 18	2. 48
3. 72	4. 64
5. 124	6. 31
7. $24a^4b^6$	8. $36x^6y^4$

Name_____ Date _____

Find the greatest common factors of the numbers or terms below

1. 15 and 18

2. 24 and 32

3. 9 and 36

4. 12 and 27

5. 27 and 162

6. 18 and 54

7. a^6 and a^{10}

8. x^4 and x^6

Name_____ **Date** _____

Find the least common multiple of the given numbers or terms below.

1. 3 and 5	2. 6 and 8
3. 4 and 12	4. 6 and 14
5. 4 , 6, and 12	6. . 6, 8, and 16
7. $4a^4$ and $6a^6$	8. $6a^4x^6$ and $24a^8x^4$

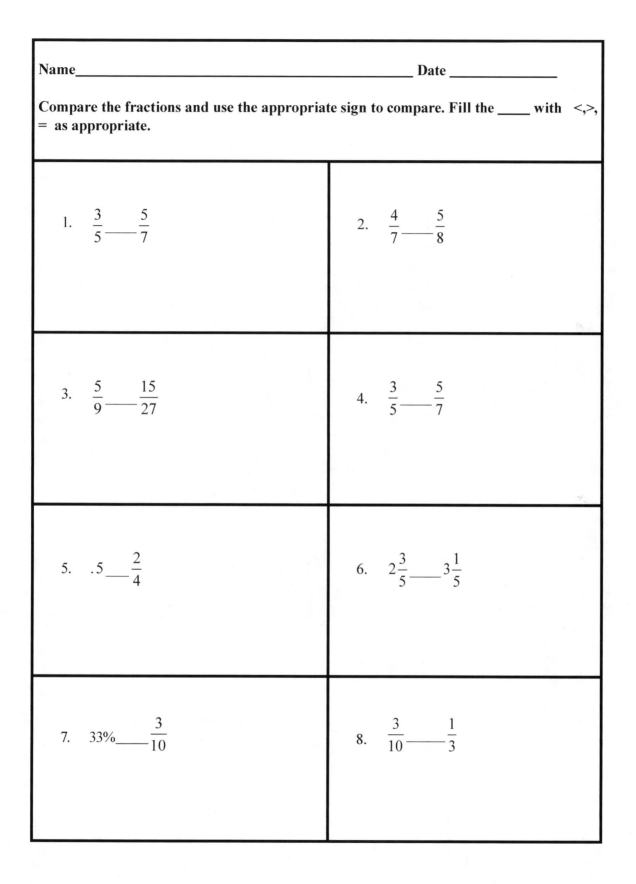

Name_____ Date _____

Compare the fractions and use the appropriate sign to compare. Fill the _____ with <,>, = as appropriate.

1. $\dfrac{3}{5}$ ___ $\dfrac{5}{7}$

2. $\dfrac{4}{7}$ ___ $\dfrac{5}{8}$

3. $\dfrac{5}{9}$ ___ $\dfrac{15}{27}$

4. $\dfrac{3}{5}$ ___ $\dfrac{5}{7}$

5. $.5$ ___ $\dfrac{2}{4}$

6. $2\dfrac{3}{5}$ ___ $3\dfrac{1}{5}$

7. 33% ___ $\dfrac{3}{10}$

8. $\dfrac{3}{10}$ ___ $\dfrac{1}{3}$

Name_____ **Date** _____

Change repeating fractions to fraction

1. X=0.222222222………..

2. X=1.333333333…….……..

3. X= 0.555555………….…..

4. X=0.25252525………

5. X=0.121212……...

6. 0.123123123123………….

7. X=1.02222222222………...

8. X= 2.033333333…………

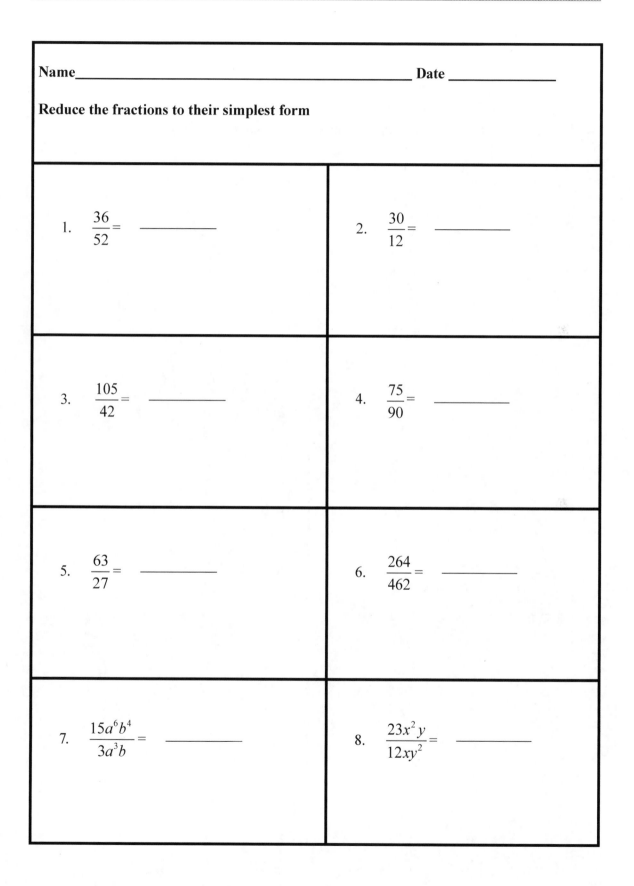

Name_____ Date _____

Reduce the fractions to their simplest form

1. $\dfrac{36}{52} =$ _____

2. $\dfrac{30}{12} =$ _____

3. $\dfrac{105}{42} =$ _____

4. $\dfrac{75}{90} =$ _____

5. $\dfrac{63}{27} =$ _____

6. $\dfrac{264}{462} =$ _____

7. $\dfrac{15a^6 b^4}{3a^3 b} =$ _____

8. $\dfrac{23x^2 y}{12xy^2} =$ _____

Name_____ Date _____

Find a. reciprocal for questions 1 to 3

      b. change to mixed fraction for question 4-6, and change to regular fraction for 7 and 8

1. $\dfrac{5}{7}$

2. $\dfrac{1}{7}$

3.   5

4. $\dfrac{23}{4}$

5. $\dfrac{27}{7}$

6. $\dfrac{28}{5}$

7. $5\dfrac{2}{3}$

8. $3\dfrac{2}{7}$

Name_____ **Date** _____

Find the sum of the given fractions.

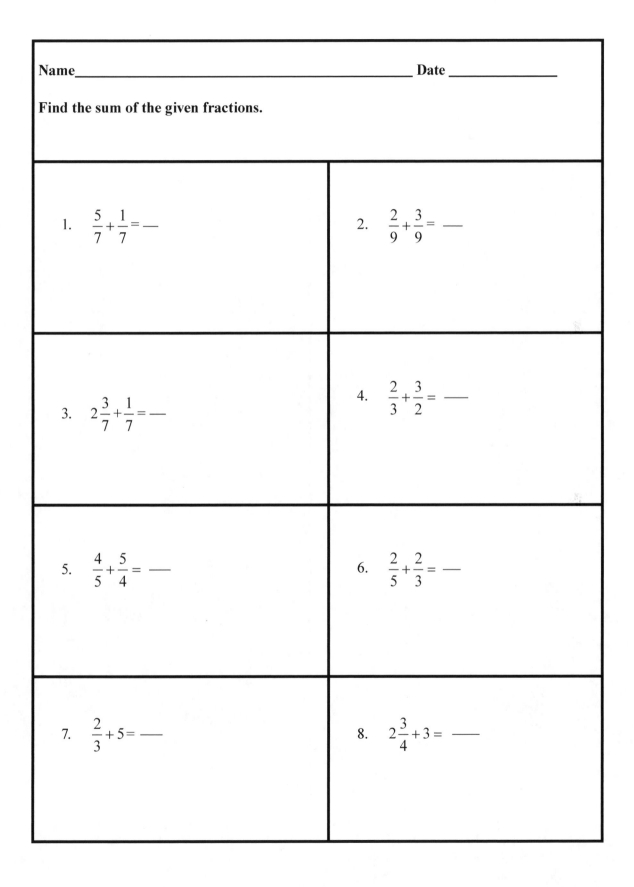

1. $\dfrac{5}{7}+\dfrac{1}{7}=$ —

2. $\dfrac{2}{9}+\dfrac{3}{9}=$ —

3. $2\dfrac{3}{7}+\dfrac{1}{7}=$ —

4. $\dfrac{2}{3}+\dfrac{3}{2}=$ —

5. $\dfrac{4}{5}+\dfrac{5}{4}=$ —

6. $\dfrac{2}{5}+\dfrac{2}{3}=$ —

7. $\dfrac{2}{3}+5=$ —

8. $2\dfrac{3}{4}+3=$ —

Name_____ Date _____

Subtract the given fractions.

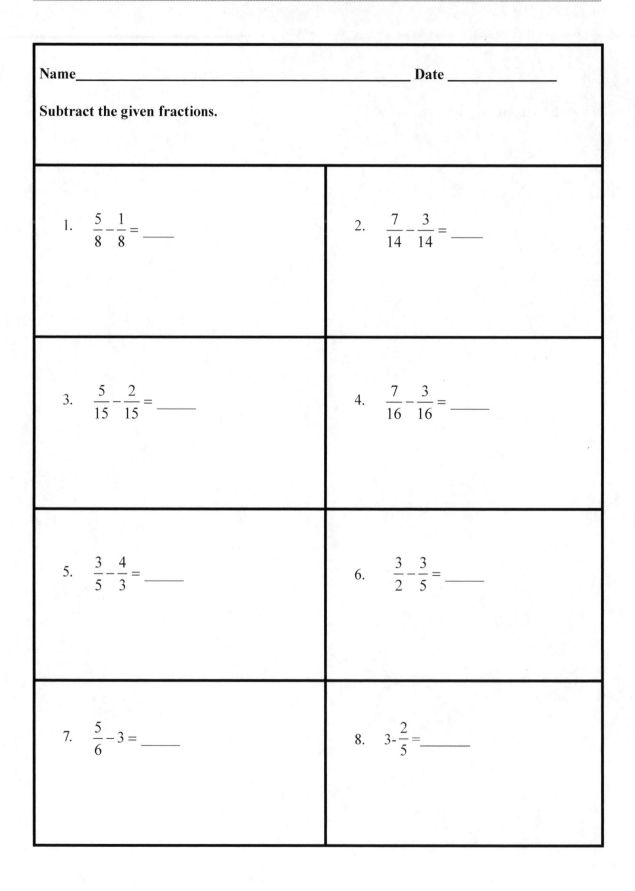

1. $\dfrac{5}{8} - \dfrac{1}{8} =$ _____

2. $\dfrac{7}{14} - \dfrac{3}{14} =$ _____

3. $\dfrac{5}{15} - \dfrac{2}{15} =$ _____

4. $\dfrac{7}{16} - \dfrac{3}{16} =$ _____

5. $\dfrac{3}{5} - \dfrac{4}{3} =$ _____

6. $\dfrac{3}{2} - \dfrac{3}{5} =$ _____

7. $\dfrac{5}{6} - 3 =$ _____

8. $3 - \dfrac{2}{5} =$ _____

Name_____ Date _____

Multiply the given fractions below.

1. $\frac{2}{5} \times \frac{3}{4} =$ _____

2. $\frac{3}{7} \times \frac{7}{3} =$ _____

3. $\frac{2}{7} \times 3 =$ _____

4. $3 \times \frac{2}{7} =$ _____

5. $2\frac{3}{4} \times \frac{3}{2} =$ _____

6. $0.5 \times \frac{2}{3} =$ _____

7. $\frac{4}{7} \times \frac{7}{2} =$ _____

8. $0.7 \times \frac{5}{2} =$ _____

Name_____ **Date** _____

Divide the questions below.

1. $\dfrac{3}{4} \div \dfrac{3}{4} =$ _____

2. $3 \div \dfrac{2}{9} =$ _____

3. $\dfrac{2}{9} \div 3 =$ _____

4. $3\dfrac{1}{3} \div \dfrac{4}{5} =$ _____

5. $0.3 \div \dfrac{4}{9} =$ _____

6. $\dfrac{4}{9} \div 0.3 =$ _____

7. $\dfrac{a}{b} \div \dfrac{a}{b} =$ _____

8. $\dfrac{a}{b} \div \dfrac{b}{a} =$ _____

Name_____ Date _____

Write each fraction to decimal. (Round your answer to the thousands)

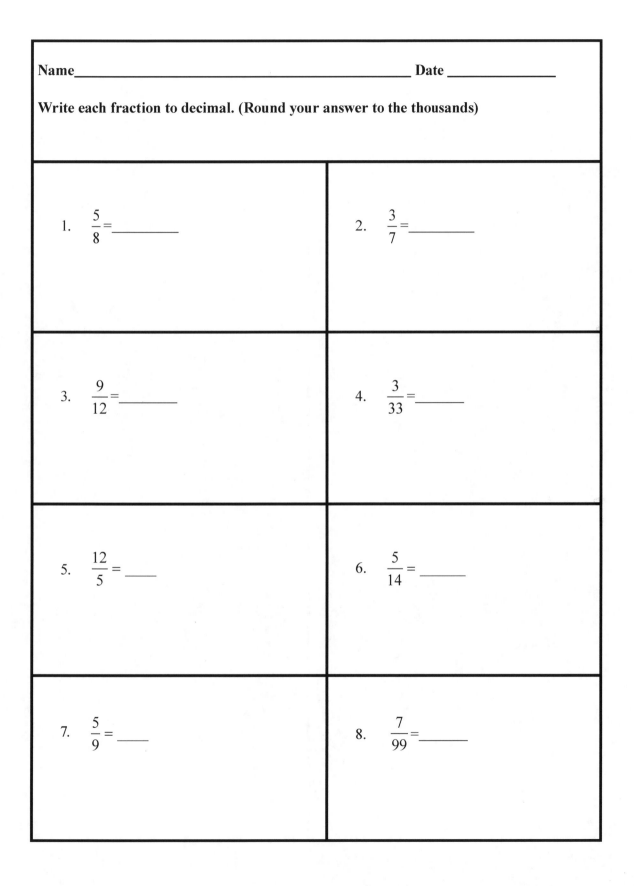

1. $\dfrac{5}{8} =$ _____

2. $\dfrac{3}{7} =$ _____

3. $\dfrac{9}{12} =$ _____

4. $\dfrac{3}{33} =$ _____

5. $\dfrac{12}{5} =$ _____

6. $\dfrac{5}{14} =$ _____

7. $\dfrac{5}{9} =$ _____

8. $\dfrac{7}{99} =$ _____

Name_____ **Date** _____

Write each decimal as fraction.

1. 0.5 = _____

2. 0.05 =_____

3. 1.05 =_____

4. 0.356 =_____

5. 0.00356 = _____

6. 23.005

7. 345.2

8. 2.031=_____

Name_____ **Date** _____

Order each set of numbers from least to greatest.

1.   0.5, 0.05 , 0.55, 0.51, 0.0055

_____, _____, _____, _____, _____
least   **greatest**

2.   0.33, $\dfrac{1}{3}$, 0.0333, $\dfrac{3}{10}$, $\dfrac{333}{1000}$

_____, _____, _____, _____, _____
least   **greatest**

3.   0.26, 0.62, 0.072, 0.56, -1.2

_____, _____, _____, _____, _____
least   **greatest**

4.   -0.9, -0.89, -.09, -1.2, -0.095

_____, _____, _____, _____, _____
least   **greatest**

5.   0.35,-0.35,0.65,-0.65,.056

_____, _____, _____, _____, _____
least   **greatest**

6. $\dfrac{3}{4}, \dfrac{4}{5}, \dfrac{5}{6}, \dfrac{6}{7}, \dfrac{7}{8}$

_____, _____, _____, _____, _____
least   **greatest**

7.   -2.3, -2.5, -3.2, -0.35, 0.30

_____, _____, _____, _____, _____
least   **greatest**

8.   0.85, 0.085, 1.58, 0.8,0.80

_____, _____, _____, _____, _____
least   **greatest**

Name_____ Date _____

Write each fraction or decimal as a percent.(round to the tenth.)

1. $\frac{3}{4}$ = _____ %

2. $\frac{5}{8}$ = _____ %

3.   0.35=_____ %

4.   0.453 =_____ %

5.   1.456=_____ %

6. $2\frac{3}{4}$ = _____ %

7.   5.368=_____ %

8.   12.35=_____%

Name_____ **Date** _____

Find each value. (Round your answer to the nearest tenth if needed)

1. 50% of 70 is what number?

2. What number is 70% of 90?

3. 15 is 20% of what number?

4. 35% of 150 is what number?

5. What percent of 80 is 30?

6. What percent of 49 is 7?

7. 20% of 70% of 120 is what number?

8. 40 is what percent of 120?

Name_____ Date _____

Add the given decimals

1.　　$35.3 + 23.51 =$ _____

2.　　$123.05 + 2.31 =$ _____

3.　　$1.035 + .0132 =$ _____

4.　　$0.103 + 321.0152 =$ _____

5.　　$12.031 + 3.012 =$ _____

6.　　$.1325 + .2 =$ _____

7.　　$7 + .0256 =$ _____

8.　　$12.031 + .102 =$ _____

Name_____ **Date** _____

Subtract the decimals

1. $1.02 - .03 =$ _____

2. $2.03 - 1.05 =$ ____

3. $3.025 - 1.05 =$ _____

4. $52.32 - 2.003 =$ _____

5. $25.132 - 3.012 =$ _____

6. $3.12 - 1.02 =$ _____

7. $42.132 - 23.01 =$ _____

8. $43.23 - 24.3 =$ _____

Name_____ **Date** _____

Multiply the given decimals

1. $2 \times 3.2 =$ _____

2. $2.3 \times 3.2 =$ _____

3. $2.03 \times 3.2 =$ _____

4. $2.03 \times 3.02 =$ _____

5. $0.23 \times 0.32 =$ _____

6. $0.023 \times 0.032 =$ _____

7. $4.023 \times 0.2 =$ _____

8. $0.0001 \times 0.0001 =$ _____

Name_____ Date _____

Divide the given decimals. Round to the nearest tenth if necessary.

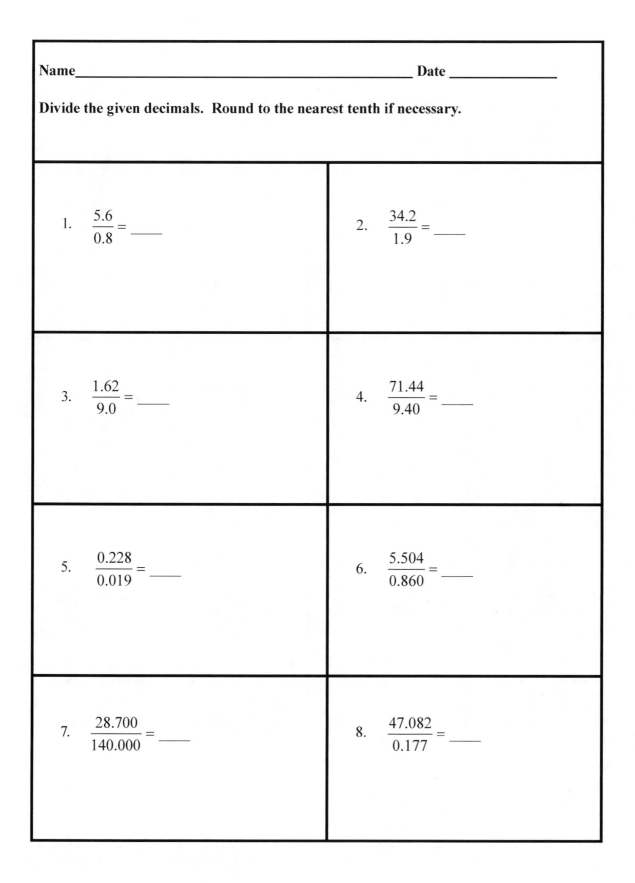

1. $\dfrac{5.6}{0.8} =$ ____

2. $\dfrac{34.2}{1.9} =$ ____

3. $\dfrac{1.62}{9.0} =$ ____

4. $\dfrac{71.44}{9.40} =$ ____

5. $\dfrac{0.228}{0.019} =$ ____

6. $\dfrac{5.504}{0.860} =$ ____

7. $\dfrac{28.700}{140.000} =$ ____

8. $\dfrac{47.082}{0.177} =$ ____

Name_____ Date _____

For questions 1-4 determine if the given measures can be the lengths of sides of a triangle and write yes or no. For questions 5-8 find the range of the third side of a triangle given the measures of the two sides.

1. 4,8, and 11

2. 14,17, and 31

3. 6,7, and 12

4. 9,10, and 20

5. 8, and 13

6. 15, and 17

7. 7, and 15

8. 10, and 14

Name_____ Date _____

Simplify without using calculator.

1. $\sqrt{4} = $ ____

2. $\sqrt{8} = $ ____

3. $\sqrt{20} = $ ____

4. $\sqrt{108} = $ ____

5. $\sqrt{6} \times \sqrt{24}$

6. $\sqrt{275} = $ ____

7. $\dfrac{\sqrt{81}}{\sqrt{27}} = $ ____

8. $\sqrt{\dfrac{3}{4}} \times \sqrt{\dfrac{4}{5}} = $ ____

Name_____ Date _____

Simplify without calculator.

1. $\sqrt{2} + 3\sqrt{2} =$ ___

2. $5\sqrt{3} - 3\sqrt{3} =$ _____

3. $7\sqrt{5} + 12\sqrt{5} =$ _____

4. $4\sqrt{3} + 2\sqrt{12} =$ _____

5. $\sqrt{27} + \sqrt{48} + \sqrt{12} =$ _____

6. $\sqrt{72} + \sqrt{50} - \sqrt{8} =$ _____

7. $\sqrt{90} + \sqrt{10} - \sqrt{40} =$ _____

8. $\sqrt{\sqrt[3]{64}} =$ _____

Name_____ Date _____

Rationalize the denominator.

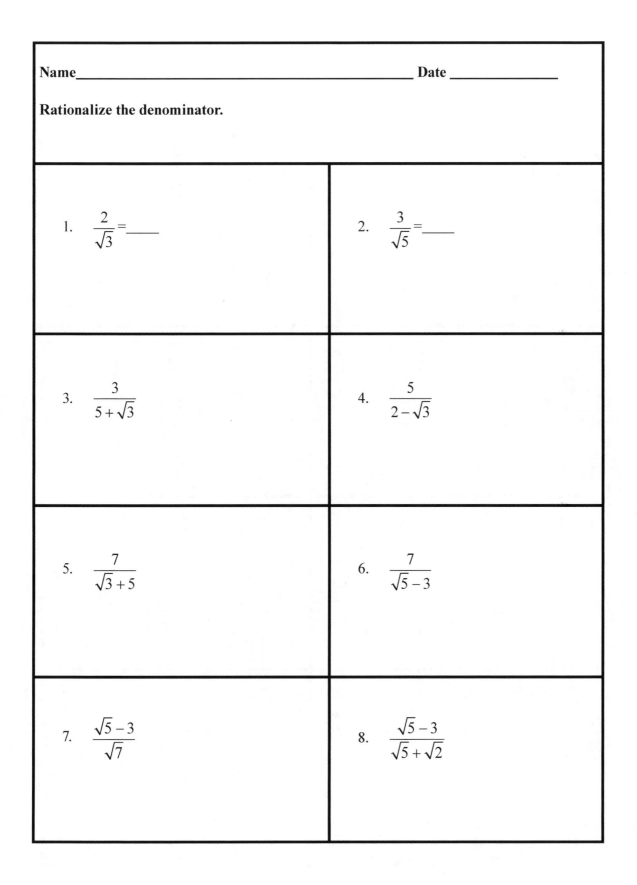

1. $\dfrac{2}{\sqrt{3}} =$ _____

2. $\dfrac{3}{\sqrt{5}} =$ _____

3. $\dfrac{3}{5+\sqrt{3}}$

4. $\dfrac{5}{2-\sqrt{3}}$

5. $\dfrac{7}{\sqrt{3}+5}$

6. $\dfrac{7}{\sqrt{5}-3}$

7. $\dfrac{\sqrt{5}-3}{\sqrt{7}}$

8. $\dfrac{\sqrt{5}-3}{\sqrt{5}+\sqrt{2}}$

Name_____ **Date** _____

Factor out the common terms

1. $8 + 16a$

Common factor=_____

2. $12a^2 + 18a$

Common factor=_____

3. $8a^2b + 12a^3b^2c$

Common factor=_____

4. $8ab - 12a^2b + 18ab^2$

Common factor=_____

5. $12x^3y - 6xy^2 + 4xy$

Common factor=_____

6. $9abc + 12bc + 4c$

Common factor=_____

7. $15p^2 - 36pq$

Common factor=_____

8.   12a-6

Common factor=_____

Name_____ Date _____

Solve each proportion.

1. $\dfrac{1}{5} = \dfrac{x}{15}$

2. $\dfrac{2x}{15} = \dfrac{2}{3}$ –

3. $\dfrac{8}{3x} = \dfrac{4}{6}$

4. $\dfrac{x+5}{3} = \dfrac{2}{3}$

5. $\dfrac{2x-1}{x+2} = \dfrac{5}{2}$

6. $\dfrac{x-5}{2x} = \dfrac{2}{7}$

7. $\dfrac{2x-5}{x+2} = \dfrac{3}{2}$

8. $\dfrac{3x-1}{2x-3} = 1$

Name_____ Date _____

Solve each equation

1. $x + 7 = 12$

2. $2x + 5 = 15$

3. $3x - 3 = 9$

4. $2x - 4 = 5x - 16$

5. $\dfrac{x}{5} = 2$

6. $\dfrac{3x}{5} - 1 = 2$

7. $\dfrac{x + 3}{2} = 2x$

8. $3(2x - 1) - 4 = x + 3$

Name_____ **Date** _____

Find the slope of a line that passes through the pair of points.

1. (2,3), and (6,11)

2. (1,3), and (3,9)

3. (1,5), and (2,1)

4. (-3,2), and (1,6)

5. (4,6), and (4,9)

6. (6,7), and (9,7)

7. (1,5), and (3,0)

8. (7,2), and (9,1)

Name_____ **Date** _____

Write an equation of a line that passes through the given point and is parallel to the graph of each equation.

1. $y = 2x - 4$; (2,3)

2. $y = 3x - 1$; (-1,-2)

3. $y = 2x - 5$; (1,3)

4. $y = 4x + 3$; (1-3)

5. $y = 7x - 4$; (3,-1)

6. $y = 3x + 4$; (2,3)

7. $y = 6x - 5$; (1,-1)

8. $y = 5x + 2$; (2,4)

Name_____ **Date** _____

Find the solution for the problems

---

1. $x + 6y = 18$

   $2x - 3y = -24$

2. $2x - 7y = -7$

   $x + 14y = 84$

---

3. $x - 3y = -4$

   $2x + 6y = 5$

4. $5x + 8y = 28$

   $7x + 4y = -4$

---

5. $4x - 2y = -14$

   $3x - y = -8$

6. $5x - 2y = -10$

   $3x + 6y = 66$

---

7. $3x - 4y = 3$

   $4x - 3y = 11$

8. $3x - 2y = 1$

   $x + 4y = 9$

Name_____ Date _____T90Awp-10

Translate the problems into algebraic statements

1. Twice a number

2. Five more than a number

3. Five more than three times a number.

4. A number increased by four.

5. Five times a number increased by four.

6. Two consecutive numbers.

7. Two consecutive even integers.

8. Two consecutive odd integers.

Name_____ **Date** _____

Find the solution for the given word problems.

1. If Hanna cleans the room, it takes her 7 minutes. The same job takes 5 minutes for her older sister. How long will it take if both work together?

2. If Hanna cleans the room, it takes her 7 minutes. Her friend came one day to help and it took 3 minutes for both to finish. How long will it take for the friend if she worked by herself?

3. The sum of the ages of Hanna and her father is 50. Five years from now, the father will be four times as old as his daughter. How old are the daughter and father now?

4. The sum of the ages of Alex and his father is 38. Five years ago, the father was four times as old as his son. How old are the father and son now?

5. The sum of two numbers is 25. If the first number is four times the second, what are the numbers?

6. The sum of two consecutive even integers is 30. What are the numbers?

7. The sum of an odd integer and five times the next odd integer is 52. What are the two numbers?

8. Hanna has scores of 85, 86, and 87 on three tests. What must she score on the next test to get a grade of 90?

Name_____ **Date** _____T100A-2

Solve the given absolute value problems

1. $|2x-5|=11$

2. $|5x-7|\le 11$

3. $5+|2x-5|=15$

4. $5|2x-5|=15$

5. $3+5|3x-5|\ge 18$

6. $|2x-5|=-3$

7. $-3|3x-4|=-9$

8. $|5-3x|\le 16$

Name_____ Date _____

Find the inverse of the given functions below.

1. $f(x) = 2x - 4$

2. $g(x) = 5x - 3$

3. $h(x) = \dfrac{x+2}{4}$

4. $f(x) = 3 - 5x$

5. $f(x) = 7 - 5x$

6. $f(x) = \dfrac{2 - 3x}{4}$

7. *If* g(x)=3x-2; then find $g(g^{-1}(x)) =$

8. $f(x) = \dfrac{1}{5}x - 4$

Name_____ Date _____

Evaluate the piecewise function at the given points.

1. $f(x) = \begin{cases} 2x+2 & \text{if } x \leq -2 \\ 5x-3 & \text{if } x > -2 \end{cases}$

Evaluate

    a when x=-5

    b when x=3

2. $f(x) = \begin{cases} 2-5x & \text{if } x \leq -3 \\ x & \text{if } -3 \leq x \leq 2 \\ 5 & \text{if } x > 2 \end{cases}$

Evaluate

    a. when x=-5

    b. when x=-1

    c. when x=6

3. $f(s) = \begin{cases} x^2 & \text{if } x < 0 \\ 5 & \text{if } x \geq 0 \end{cases}$

Evaluate

    a. *when* x=-3

    b.when x=3

4. $g(x) \begin{cases} x^3 & \text{if } x \leq 2 \\ 2-x^2 & \text{if } 2 < x \leq 7 \\ 5 & x > 7 \end{cases}$

Evaluate

    a. when x=0

    b. when x=5

    c. when x=9

5. $h(x) = \begin{cases} 5x^2-4 & \text{if } x \leq -2 \\ 2-x^2 & \text{if } x > -2 \end{cases}$

Evaluate

    a. when x=-4

    b. when x=2

5. $h(x) = \begin{cases} 5x^2-4 & \text{if } x \leq -2 \\ 2-x^2 & \text{if } x > -2 \end{cases}$

Evaluate

    a. when x=-4

    b. when x=2

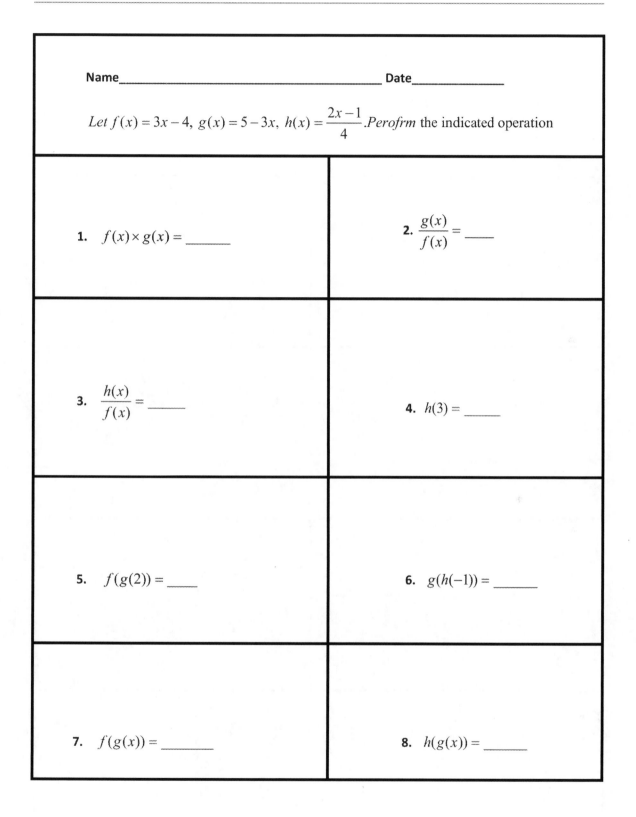

Name_____ Date_____

Let $f(x) = 3x - 4$, $g(x) = 5 - 3x$, $h(x) = \dfrac{2x - 1}{4}$. *Perofrm* the indicated operation

1. $f(x) \times g(x) = $ _____

2. $\dfrac{g(x)}{f(x)} = $ _____

3. $\dfrac{h(x)}{f(x)} = $ _____

4. $h(3) = $ _____

5. $f(g(2)) = $ _____

6. $g(h(-1)) = $ _____

7. $f(g(x)) = $ _____

8. $h(g(x)) = $ _____

Name_____ **Date** _____

Find the absolute value of the complex number.

1. $4-3i$	2. $3-5i$
3. $7-6i$	4. $-3-7i$
5. $5+i\sqrt{3}$	6. $\sqrt{5}+2i$
7. $\sqrt{2}+i\sqrt{5}$	**8.** $-\sqrt{3}-i\sqrt{7}$

Name_____ Date _____

Write the expression as a complex number in a standard form.

1. $\dfrac{5}{7-5i}$

2. $\dfrac{6-1}{3-2i}$

3. $\dfrac{7+4i}{-9-6}$

4. $\dfrac{3-i}{5+2i}$

5. $\dfrac{2+4i}{\sqrt{2}-i}$

6. $\dfrac{2+9i}{\sqrt{3}-i}$

7. $\dfrac{3+6i}{-7-2}+4-i$

8. $\dfrac{9}{4-2i}-(6i+1)$

Name_____ Date _____

Write the expression as a complex number in a standard form.

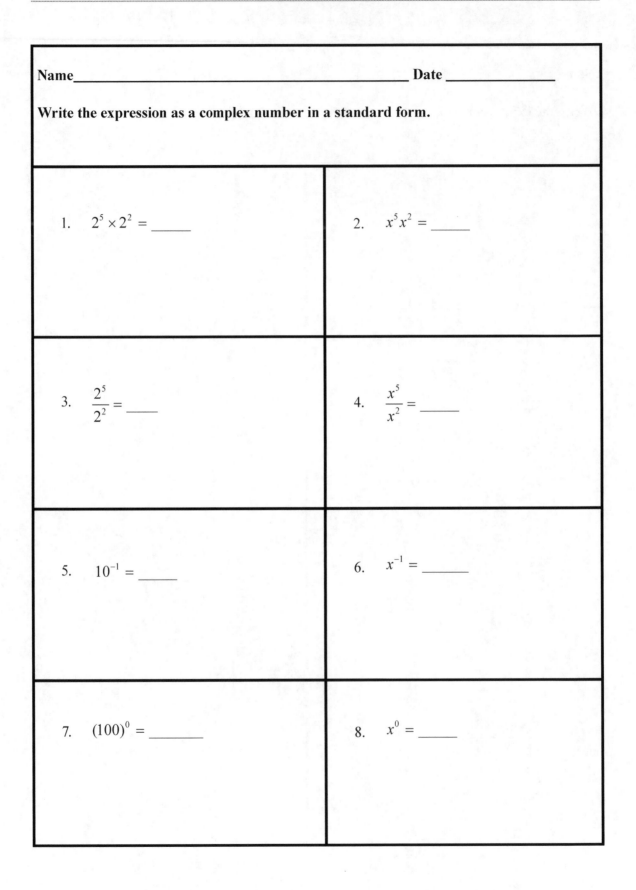

1. $2^5 \times 2^2 = $ _____

2. $x^5 x^2 = $ _____

3. $\dfrac{2^5}{2^2} = $ _____

4. $\dfrac{x^5}{x^2} = $ _____

5. $10^{-1} = $ _____

6. $x^{-1} = $ _____

7. $(100)^0 = $ _____

8. $x^0 = $ _____

Name_____ Date _____

Simplify the problems using laws of exponent

1. $(2^3)^2 =$ _____

2. $(x^3)^2 =$ _____

3. $(2 \times 3)^2 =$ _____

4. $(x \times y)^2 =$ _____

5. $\left(\dfrac{3}{6}\right)^{-1} =$ _____

6. $\left(\dfrac{x}{y}\right)^{-1} =$ _____

7. $\left(\dfrac{(4a^3b^2c)^2}{(8a^6b^8c^5)^3}\right)^{-3} =$ _____

8. $\left(\dfrac{(8x^3yz^4)^2}{(2xy^3)}\right)^{3} =$ _____

Name_____ Date _____

Rewrite the expression using rational exponent notation.

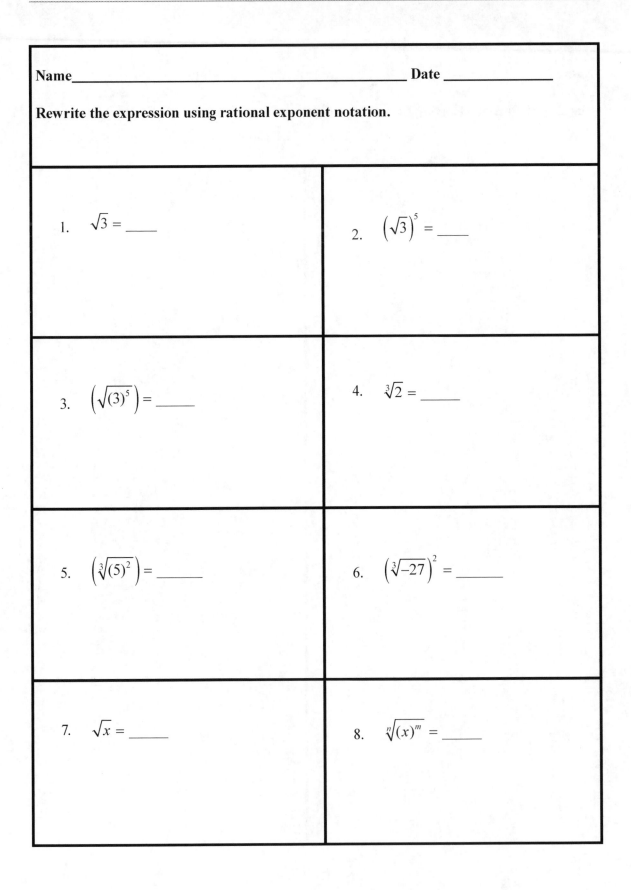

1. $\sqrt{3} =$ _____

2. $\left(\sqrt{3}\right)^5 =$ _____

3. $\left(\sqrt{(3)^5}\right) =$ _____

4. $\sqrt[3]{2} =$ _____

5. $\left(\sqrt[3]{(5)^2}\right) =$ _____

6. $\left(\sqrt[3]{-27}\right)^2 =$ _____

7. $\sqrt{x} =$ _____

8. $\sqrt[n]{(x)^m} =$ _____

Name_____ Date _____

Find the real number solution , if it exists, for the following problems

1. $x^2 - 4 = 0$	2. $x^2 + 4 = 0$
3. $x^2 + 2x + 1 = 0$	4. $x^2 - 2x + 1 = 0$
5. $x^2 + 5x + 4 = 0$	6. $x^2 - 5x + 4 = 0$
7. $x^2 + 7x - 8 = 0$	**8.** $x^2 + 8x - 9 = 0$

Name_____ **Date** _____

Factor the expression by using rule of difference of two squares

1. $x^2 - y^2$	2. $3^2 - 2^2$
3. $a^2 - 4b^2$	4. $4x^2 - 9y^2$
5. $9x^2 - 16y^2$	6. $9x^2 - y^2$
7. $(x-y)^2 - (x+y)^2$	8. $(3x-1)^2 - (2x+1)^2$

Name_____ **Date** _____

Factor the expression

1. $x^2 + x - 6$

2. $x^2 + x - 2$

3. $x^2 - 2x + 1$

4. $x^2 - 2x - 3$

5. $x^2 - 5x - 6$

6. $x^2 - 7x - 8$

7. $x^2 - 6x - 7$

8. $x^2 - 9x + 8$

Name_____ Date _____

Factor the expression

1. $x^2 - 2x$

2. $2x^2 - 8$

3. $2x^2 - 3x$

4. $5x^2 - 20x$

5. $2x^3 - 8x$

6. $x^3 - x$

7. $4x^2 - 25y^2$

8. $a^3 - b^3$

Name_____ Date _____

Solve the equation.

---

1. $5x^2 - 10x = 0$

2. $3x^2 - 12x = 0$

---

3. $3x^2 - 12 = 0$

4. $2x^2 - x - 3 = 0$

---

5. $3x^2 - 5x - 2 = 0$

6. $5x^2 - x - 6 = 0$

---

7. $7x^2 - x - 6 = 0$

8. $3x^2 = x + 2$

Name_____ **Date** _____

Find the discriminant and write the type and number of solution(s).

1. $5x^2 - 2x + 3 = 0$

2. $2x^2 + 7x + 1 = 0$

3. $3x^2 + 6x + 1 = 0$

4. $6x^2 + x + 10 = 0$

5. $x^2 + 4x + 4 = 0$

6. $5x^2 + 6x = 0$

7. $5x^2 + 6 = 0$

8. $3x^2 - 9 = 0$

Name_____ **Date** _____

Find the product

1. $(x+5)(x-2)$

2. $(x+2)(x+2)$

3. $(x-4)(x-3)$

4. $(x-1)(x-3)$

5. $(2x-3)(3x-2)$

6. $(3x-1)(x+4)$

7. $(2x-4)(3y+2)$

8. $(a-b)(c+d)$

Name_____ Date _____

Find the value of c that makes each trinomial a perfect square.

1. $(x+5)(x-2)$	2. $x^2 + 6x + c$
3. $x^2 + 8x + c$	4. $x^2 + 5x + c$
5. $x^2 + 3x + c$	6. $x^2 + \dfrac{5}{3}x + c$
7. $x^2 + \dfrac{3}{2}x + c$	8. $2x^2 - 3x + c$

Name_____ Date _____

Solve by completing the square

1. $x^2 + 6x + 2 = 0$

2. $x^2 + 4x + 2 = 0$

3. $x^2 + 3x + 4 = 0$

4. $x^2 - 6x + 4 = 0$

5. $x^2 + 5x + 1 = 0$

6. $x^2 - 5x + 2 = 0$

7. $2x^2 + 5x + 1 = 0$

8. $ax^2 + bx + c = 0$

Name_____ **Date** _____

Find a. vertex b. axis of symmetry c. maximum or minimum value
d. x and y intercepts for the given problems and sketch the graph.

1. $f(x) = (x+2)^2 + 3$

2. $f(x) = -(x-2)^2 - 3$

3. $f(x) = 3(x-5)^2 + 3$

4. $g(x) = -2(x-4)^2 + 1$

5. $f(x) = -2(x+2)^2$

6. $f(x) = -3(x-1)^2$

7. $g(x) = 5(x-1)^2$

8. $f(x) = a(x-h)^2 + k$

Name_____ Date _____

Make a table, sketch the graph, and write the domain and range for the problems below.

1. $f(x) = x$

2. $f(x) = x^2$

3. $f(x) = x^3$

4. $f(x) = |x|$

5. $f(x) = 2^x$

6. $f(x) = \log_2 x$

7. $f(x) = \sqrt{x}$

8. $f(x) = \dfrac{1}{x}$

Name_____ **Date** _____

1. *Find* the number of zeros
2. Find the maximum number of turns
3. Find the possible positive real zeros
4. Find the possible negative real zeros.
5. Find the possible immaginary zeros.
6. Even, Odd, or Neither
7. Any symmetry.

---

1. $f(x) = -3x^7 + 4x^4 + 3x^2 - 2x + 1$

2. $f(x) = 2x^4 + 7x^3 - 17x^2 - 58x - 24$

---

3. $f(x) = 5x^3 - 2x^2 + x - 3$

4. $f(x) = 5x^6 - 3x^4 + 4x + 6x^2 - x + 4$

---

5. $f(x) = 3x^6 - 2x^2 + x^3 + 2x^2 - x - 3$

6. $f(x) = x^4 + 2x^3 - 6x^2 - 3x + 2$

---

7. $f(x) = -x^6 + 2x^5 - 3x^2 + 9$

8. $f(x) = x^4 + 3x^2 - 2$

Name_____ **Date** _____

Calculate the average rate of change of the function for the given interval.

1. $f(x) = x^2 + 2x + 1$
    a. $-4 \le x \le -2$
    b. $3 \le x \le 6$

2. $f(x) = -2x^2 + x + 10$
    a. $-3 \le x \le -2$
    b. $3 \le x \le 4$

3. $f(x) = 2x^2 - x - 3$
    a. $2 \le x \le 3$
    b. $-3 \le x \le 4$

4. $f(x) = -x^2 - x$
    a. $-1 \le x \le 0$
    b. $1 \le x \le 2$

5. $f(x) = x^2$
    a. $-1 \le x \le 0$

6. $f(x) = x^2 + 3x + 2$
    a. $-3 \le x \le -1$
    b. $1 \le x \le 3$

7. $f(x) = x^2 - 1$
    a.. $1 \le x \le 3$

8. $f(x) = x^2 + 3$
    a. $-1 \le x \le 0$

Name_____ **Date** _____

Write the first six terms of the sequence

1. $a_n = n + 5$	2. $a_n = 2n - 5$
3. $a_n = 2^n - 1$	4. $a_n = \dfrac{5}{n}$
5. $a_n = 5n + 2$	6. $a_n = \dfrac{n^2}{n+1}$
7. $a_n = \dfrac{3n}{2+n}$	8. $a_n = \dfrac{2n}{7n-1}$

Name_____ Date _____

For the sequence, write the next two terms, and write the rule for the sequence.

1. $3, 7, 11, 15, 19$.............	2. $-3, -1, 1, 3, 5, 7$.................
3. $2, 5, 8, 11, 14, 17$...................	4. $14, 11, 8, 5, 2, -1, -4,$..............
5. $8, 4, 0, -4, -8, -12$................	6. $7, 4, 1, -2, -5, -8$................
7. $11, 6, 1, -4, -9, -14$...................	8. $19, 13, 7, 1, -5, -11$................

Name_____ Date _____

Find the sum of the series

1. $\displaystyle\sum_{n=1}^{5} n^2$	2. $\displaystyle\sum_{n=1}^{6} 2n - 1$
3. $\displaystyle\sum_{n=1}^{7} 3n + 2$	4. $\displaystyle\sum_{n=2}^{5} 4n + n^2$
5. $\displaystyle\sum_{n=3}^{5} 2n^2 - 4n$	6. $\displaystyle\sum_{n=1}^{3} n^3$
7. $\displaystyle\sum_{n=1}^{3} \frac{1}{n}$	8. $\displaystyle\sum_{n=2}^{4} \frac{1}{n^2}$

Name_____ **Date** _____

write a rule for the nth term of the arithmetic sequence that has the two given terms.

1. $a_2 = 2, a_{11} = 29$	2. $a_3 = 7, a_{12} = 43$
3. $a_3 = 15, a_{13} = 75$	4. $a_2 = 6, a_{10} = 30$
5. $a_5 = 27, a_{13} = 3$	6. $a_6 = 12, a_{14} = -10$
7. $a_2 = 6, a_{10} = 30$	8. $a_4 = 10, a_{14} = 25$

Name_____ **Date** _____

Write a rule for the nth term of the geometric sequesnce. Find G_4

1. $3, 6, 12, 24, 48, \ldots\ldots$	2. $5, 10, 20, 40, 80, \ldots\ldots$
3. $G_3 = 16, r = 2$	4. $G_3 = 12, r = 2$
5. $G_1 = 5, r = 3$	6. $G_1 = 9, r = 3$
7. $G_1 = 5, r = 3$	**8.** $G_1 = 3, r = 5$

Given

$$A = \begin{bmatrix} 2 & 3 \\ -5 & 4 \end{bmatrix} \qquad B = \begin{bmatrix} -1 & 2 \\ 3 & 5 \end{bmatrix} \qquad C = \begin{bmatrix} 6 & 0 \\ 2 & 7 \end{bmatrix} \qquad D = \begin{bmatrix} -3 & 2 & -1 \\ 2 & -3 & 0 \\ 4 & 1 & 3 \end{bmatrix}$$

$$E = \begin{bmatrix} 1 & 2 & -3 \\ 4 & 0 & 5 \\ 2 & -1 & 3 \end{bmatrix} \qquad F = \begin{bmatrix} 2 & 5 \\ 1 & -2 \\ 3 & 4 \end{bmatrix} \qquad G = \begin{bmatrix} 2 & -1 \\ 4 & -3 \\ -1 & 5 \end{bmatrix} \qquad H = \begin{bmatrix} 2 & -3 & 4 \\ 3 & 1 & -1 \end{bmatrix}$$

Perform the indicated operation.

1. $A + B$	2. $2B - C$
3. $C + D$	4. $4G - 2F$
5. $2C + B$	6. $5A$
7. $5D$	8. $6H$

Name_____ **Date** _____

Find the equation of a circle with the given center and radius

1. Center=(5,7) ,and radius=6	2. Center=(0,5) ,and radius =8
3. Center=(-2,-4), and radius=6	4. $Center = (0,0), and$ radius=7

Find the center and radius for the given circle #5-8 5. $(x\text{-}4)^2 + (y+6)^2 = 2^2$	6. $(x\text{-}5)^2 + (x-3)^2 = 5^2$
7. $(x\text{-}3)^2 + y^2 = 16$	8. $x^2 + y^2 = 5$

Find the equation of a circle passing through P1 and perpendicular to a line passing through p2.#9-12 9. $P_1 = (2,1)$, and (4,3) and $P_2 = (2,7)$	10. P_1=(2,3), and (5,9) and $P_2 = (3,8)$
11. $P_1 = (-5,-5), and(-1,-1)$ and $P_2 = (2,3)$	12. $P_1 = (-4,-3), and(2,3)$ and $P_2 = (5,9)$

Solve using completing the square #13-16 13. $x^2 + 6x + 1 = 0$	14. $x^2 + 3x + 5 = 0$
15. $2x^2 + 6x + 8 = 0$	16. $3x^2 + 7x + 9 = 0$

Find the equation of a line tangent to the circle $x^2 + y^2 = 10$ at (1,-3)	

Name_____	Date _____
Write the equation of a circle in standard form.	**T5-3**

1. $x^2 + y^2 + 6x + 8y + 16 = 0$	2. $x^2 + y^2 + 10x + 12y + 6 = 0$
3. $x^2 + 8x + y^2 + 6x + 15 = 0$	4. $x^2 + y^2 + 16x + 8y + 6 = 0$
5-8 Find the equation of a circle with the center and radius given. **Center=(2,5) , and radius=5**	**Center =(-3,5) and radius =6**
7. *center* (-3,-5), and radius=2	8. *center* = (-5,-3), and radius=7
$9-12$ solve by completing the square 9. $x^2 + 6x + 4 = 0$	10. $x^2 + 2x + 4 = 0$
11. $x^2 + 6x + 2 = 0$	12. $x^2 + 8x + 10 = 0$
13-14. Find the equation of a circle passing through P_1 and perpendicular for a line passing through $points$ $P_3 and$ P_4 **13.** $P_1 = (5,-1)$, and $P_2 = (-1,-3)$, $P_3 = (-5,-11)$	14. $P_1 = (-2,1), and$ $P_2 = (-1,-3)$, and $P_3 = (1,3)$
15. Find the equation of a line tangent to the circle $x^2 + y^2 = 13$ at $(2,3)$	

Name_____ **Date** _____

Solve the given problems

1. $x^2 + 6x + 7 = 0$	2. $x^2 + 4x + 6 = 0$
3. $x^2 - 6x + 2 = 0$	4. $x^2 - 4x + 2 = 0$

Write the equation of a circle in standard form

5. $x^2 + y^2 + 8x - 6y + 7 = 0$	6. $x^2 + 6x + y^2 + 12y - 2 = 0$
7. $x^2 - 8x + y^2 - 4x - 15 = 0$	8. $x^2 + y^2 - 6x + 8y + 6 = 0$

9-10 Find the equation of a circle passing through P_1 and perpendicular for a line passing through $points$ $P_3 and$ P 9. $P_1 = (3,2)$, and $P_2 = (-3,2)$, and $P_3 = (1,1)$	10. $P_1 = (4,5)$ and $P_2 = (-1,2)$, and $(3,4)$
15. Find the equation of a line tangent to the circle 11. $\begin{array}{l} x^2 + y^2 = 5 \\ at \ (1,2) \end{array}$	12. $\begin{array}{l} x^2 + y^2 = 5 \\ at \ (1,-2) \end{array}$

Name_____ Date _____

Find the equation of a parabola for the given focus and directrix.

1. *focus* (0,-3), directrix y=3	2. *focus* (0,2), directix y=-2
3. *focus* $= (3,2), directix y = 1$	4. *focus* (1,-2) directrix y=2
Write the equation of a circle in standard form 5. $x^2 + y^2 + 4x - 6y + 2 = 0$	6. $x^2 + 6x + y^2 - 4x + 8y + 3 = 0$
7. $x^2 - 10x + y^2 + 10y - 10 = 0$	8. $x^2 + y^2 - 2x + 6y + 1 = 0$
Find the equation of a circle given the center and radius 9. **Center=(5,-2) and radius =8** .	**10. center=(5,7), radius=3**
15. Find the equation of a line tangent to the circle 11. $x^2 + y^2 = 34$ *at* (3,5)	12. $x^2 + y^2 = 34$ *at* (-3,5)

Name_____ Date _____

Write the equation of each parabola in standard form.

1. *focus* (2,0), directrix y=-2	2. *focus* (0,8), directix y=-8
3. Focus (-20,0) Directrix x =20	4. *focus* (0,-1/12) directrix y=1/12
5. focus (5,5) and y=-3	6. focus (3,0) and x=-2
7. focus (4,-3) and directrix y=6	8. *focus* (8,0) and directix y=4

Find the equation of a circle given the center and radius

9. Center=(7,0) and radius =9	10. center=(6,-7), radius=9

15. Find the equation of a line tangent to the circle

11. $x^2 + y^2 = 41$ *at* (-4,5)	12. $x^2 + y^2 = 41$ *at* (-4,-5)

Name_____ Date _____T5-11

Write the equation of the hyperbola with the given foci and vertices.

1. *foci*   (6,0), (-6,0)
 	
     Vertices (4,0),(-4,0)
	2. foci (0,8),(0,-8)
	
	     Vertices (0,7),(0,-7)

3. foci (0,-8), (0,8) Vertices (0,-5),(0,5)	4. foci $(0,-2\sqrt{5}),(0,2\sqrt{5})$ Vertices (0,-4),(0,4)
5. foci (16,0),(-16,0) Vertices (14,0),(-14,0)	6. foci (0,12),(0,-12) Vertices $(0,6\sqrt{3}),(0,-6\sqrt{3})$
7. foci $(\sqrt{13},0),(-\sqrt{13},0)$ *Vertices* (2,0),(-2,0)	8. foci $(0,\sqrt{61}),(0,-\sqrt{61})$ *Vertices* (0,6),(0,-6)
9. foci (8,0),(-8,0) Vertices (6,0),(0,-6)	10. foci $(0,\frac{5}{2}),(0,-\frac{5}{2})$ *Vertices* (0,2),(0,-2)
The equation of a parabola, an ellipse, and a hyperbola are given. Find the equation. 11. $25x^2 + 4y^2 = 100$	12. $25x^2 + 4y = 0$
13. $25x^2 - 4y^2 = 100$	14. $108y^2 - 81x = 0$

Name_____ Date_____

Find the circumference or perimeter of the given shape.

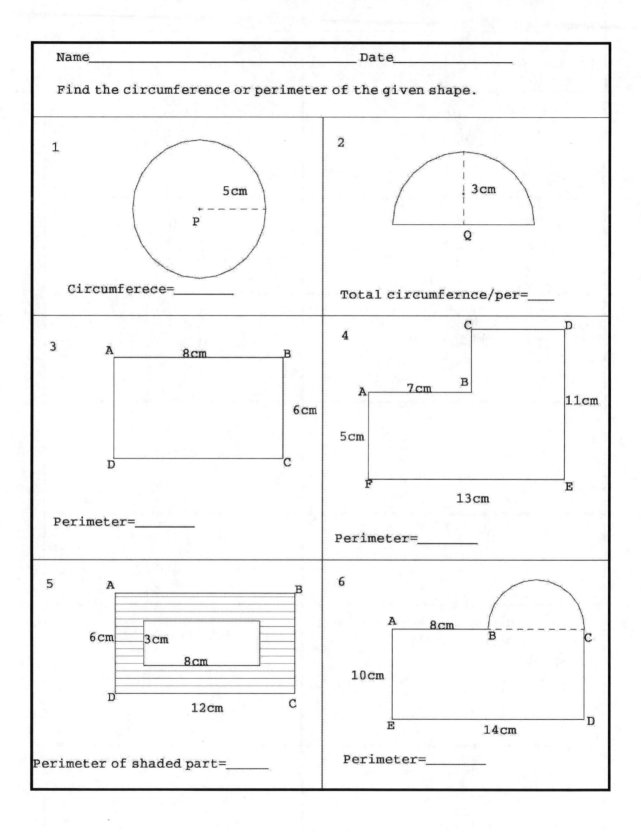

1

5cm

P

Circumferece=_____

2

3cm

Q

Total circumfernce/per=____

3

A   8cm   B

6cm

D   C

Perimeter=_____

4

C   D

7cm   B

A

11cm

5cm

F   E

13cm

Perimeter=_____

5

A   B

6cm   3cm

8cm

D   12cm   C

Perimeter of shaded part=_____

6

A   8cm   B   C

10cm

E   14cm   D

Perimeter=_____

Name_____ Date_____

Find the area of the given shapes below

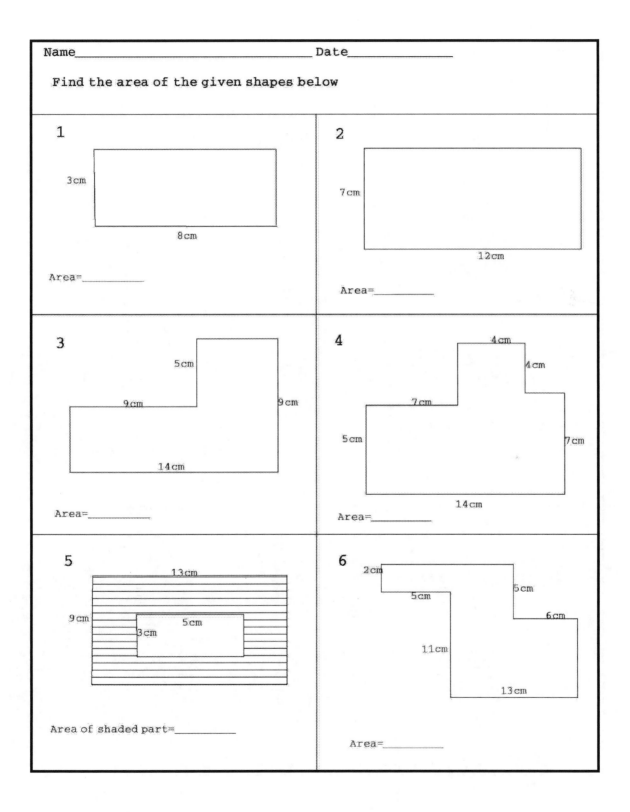

1

3cm

8cm

Area=_____

2

7cm

12cm

Area=_____

3

5cm

9cm

9cm

14cm

Area=_____

4

4cm

4cm

7cm

5cm

7cm

14cm

Area=_____

5

13cm

9cm

5cm

3cm

Area of shaded part=_____

6

2cm

5cm

5cm

6cm

11cm

13cm

Area=_____

Name_____ Date _____

Given the radius or diameter, find the area of the shapes below.

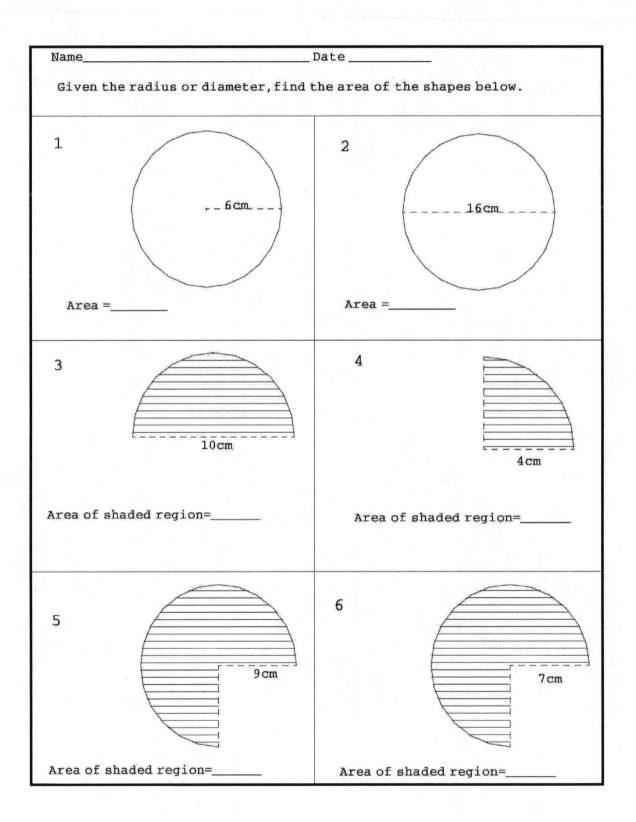

1

6cm

Area =_____

2

16cm

Area =_____

3

10cm

Area of shaded region=_____

4

4cm

Area of shaded region=_____

5

9cm

Area of shaded region=_____

6

7cm

Area of shaded region=_____

Name_____ Date_____

Find the unknown side (Round your answer to the nearest tenth)

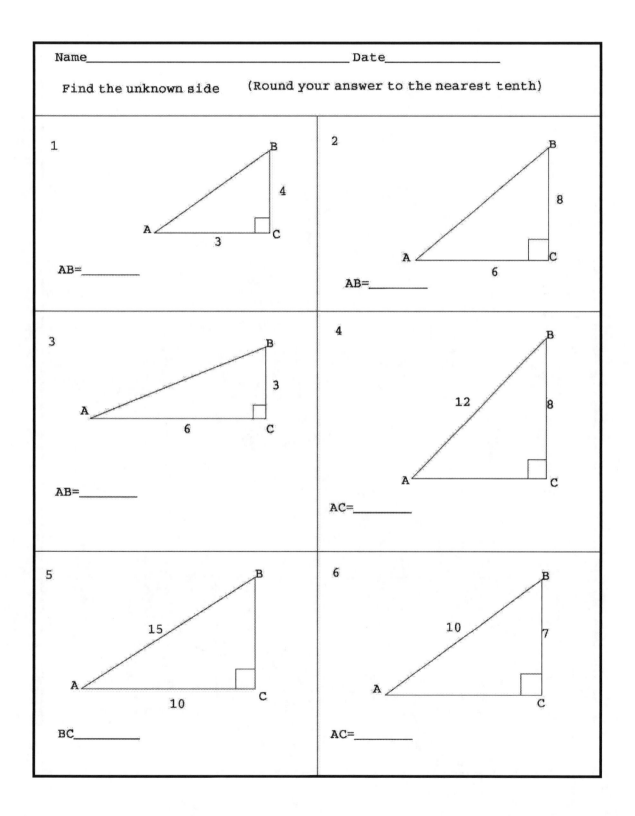

1

4

A   3   C

B

AB=_____

2

8

A   6   C

B

AB=_____

3

6   3

A   C

B

AB=_____

4

12   8

A   C

B

AC=_____

5

15

A   10   C

B

BC_____

6

10   7

A   C

B

AC=_____

Name_____ Date_____ T70GA-10

Name five different names of angles for the given shapes

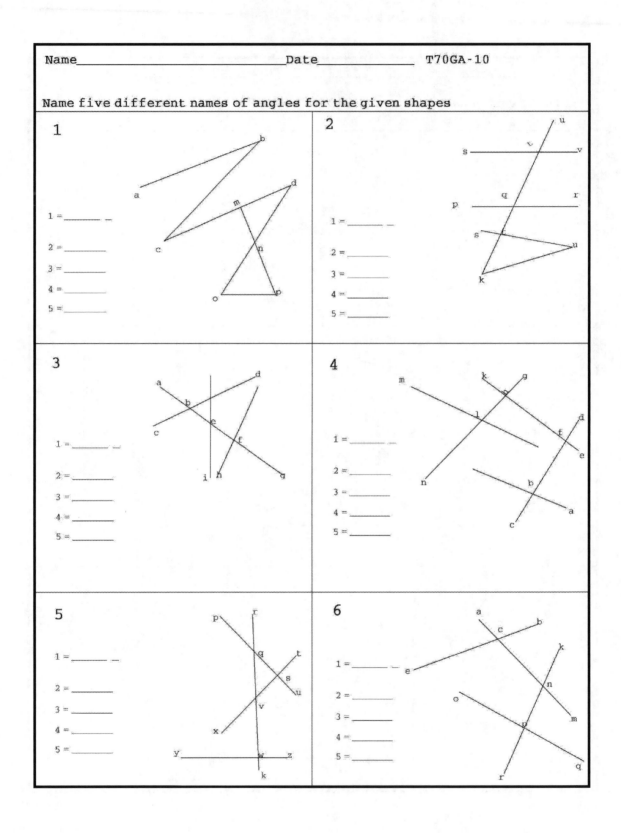

1

1 = _____

2 = _____

3 = _____

4 = _____

5 = _____

2

1 = _____

2 = _____

3 = _____

4 = _____

5 = _____

3

1 = _____

2 = _____

3 = _____

4 = _____

5 = _____

4

1 = _____

2 = _____

3 = _____

4 = _____

5 = _____

5

1 = _____

2 = _____

3 = _____

4 = _____

5 = _____

6

1 = _____

2 = _____

3 = _____

4 = _____

5 = _____

Name _____ Date_____ T70GCA-14

For the problems below, find the unknown value.

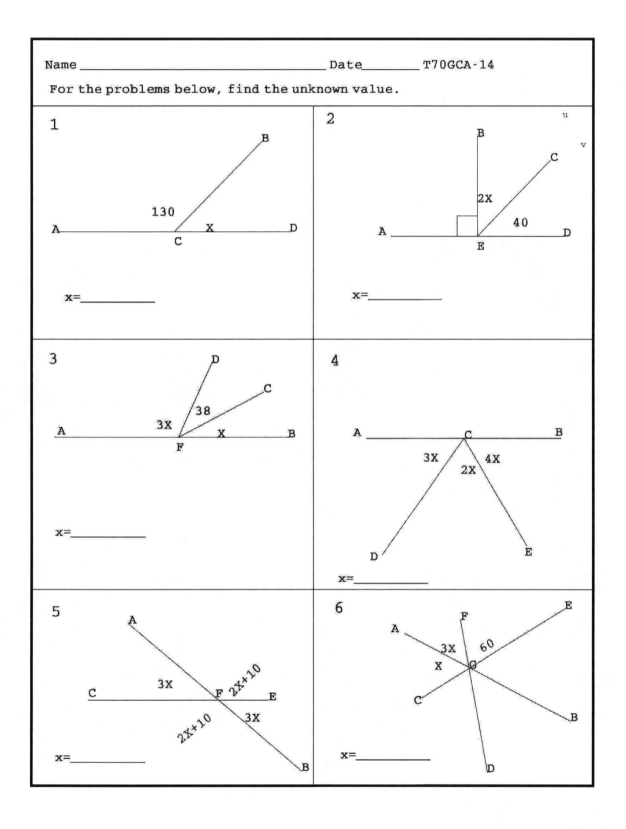

1

B

130

A   X   D
     C

x=_____

2   u

                                    B   v
                                              C
                                2X
                                        40
A _____|_____ D
              E

x=_____

3

                    D
                         C
              38
      3X   X
A _____ B
      F

x=_____

4

A _____ B
              C
      3X   2X   4X

   D   E

x=_____

5

        A

      3X   2X+10
C _____ F _____ E

      2X+10   3X

                    B

x=_____

6

                              E
              F
A   3X   60
      X   G
   C   B

              D

x=_____

73

Name_____ date_____ T50GA-11

Identify three complementary angles for each shape
(Assume line AB is perpendicular to line CD)

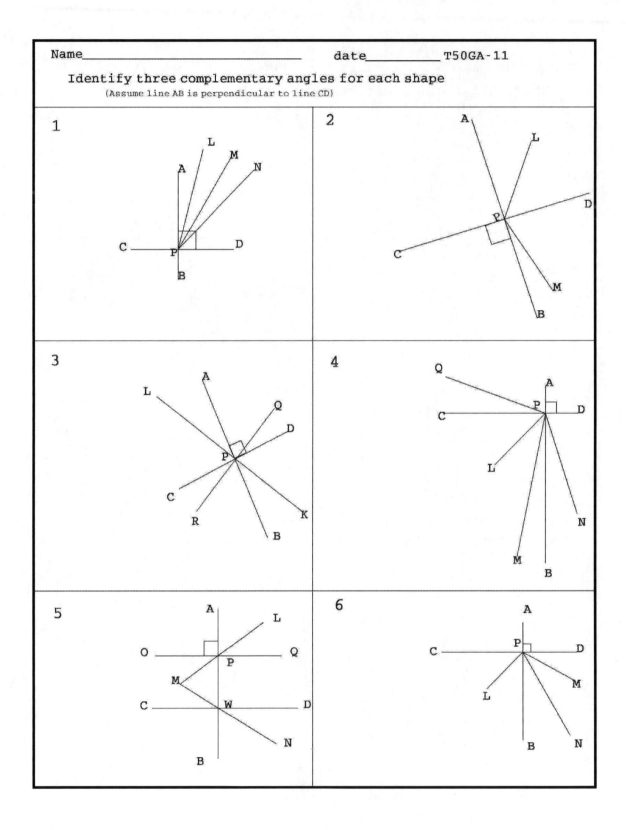

Name_____Date_____T70GCA15

Find the value of x and indicated angle.

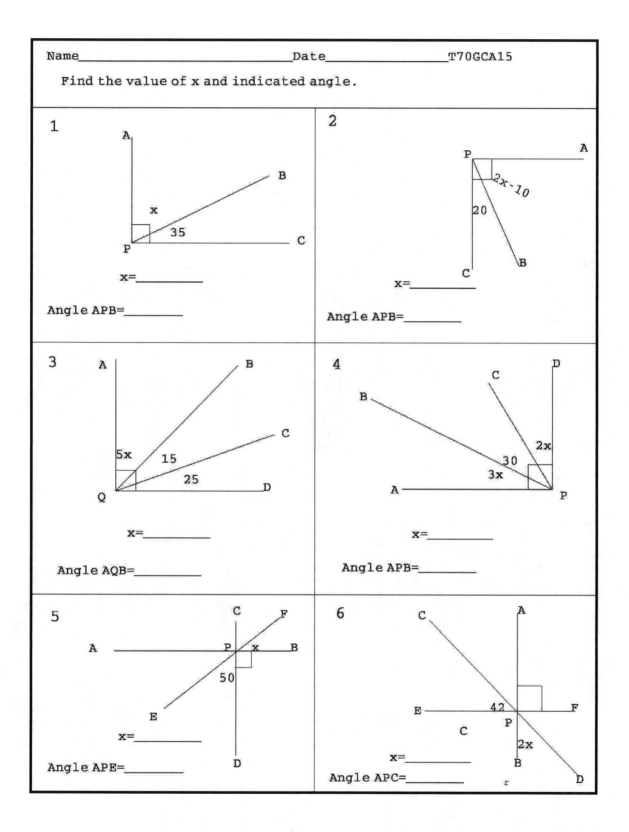

1

A

B

x

35

P

C

x=_____

Angle APB=_____

2

P

A

2x-10

20

C

B

x=_____

Angle APB=_____

3

A

B

5x

15

C

25

D

Q

x=_____

Angle AQB=_____

4

C

D

B

2x

30

3x

A

P

x=_____

Angle APB=_____

5

C

F

A

P

x

B

50

E

D

x=_____

Angle APE=_____

6

C

A

E

42

F

C

P

2x

B

D

x=_____

Angle APC=_____

Name_____ Date_____

List the sides in order from shortest to longest

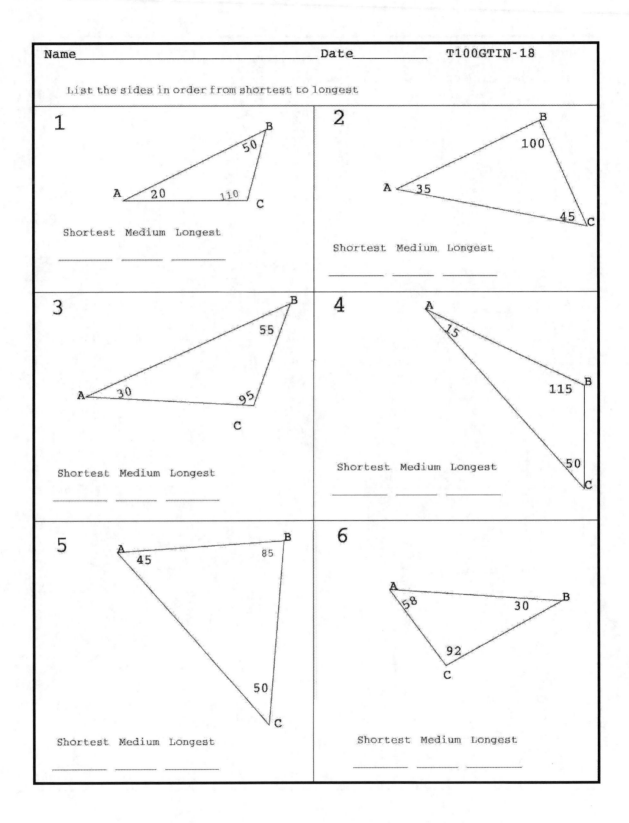

1

B
50
A 20   110 C

Shortest Medium Longest

_____ _____ _____

2

B
100
A 35
45 C

Shortest Medium Longest

_____ _____ _____

3

B
55
A 30
95
C

Shortest Medium Longest

_____ _____ _____

4

A
15
115 B
50
C

Shortest Medium Longest

_____ _____ _____

5

A 45   85 B

50
C

Shortest Medium Longest

_____ _____ _____

6

A
58   30   B
92
C

Shortest Medium Longest

_____ _____ _____

Name_____ Date_____ T100GTAIN-19

List the angles from smallest to largest

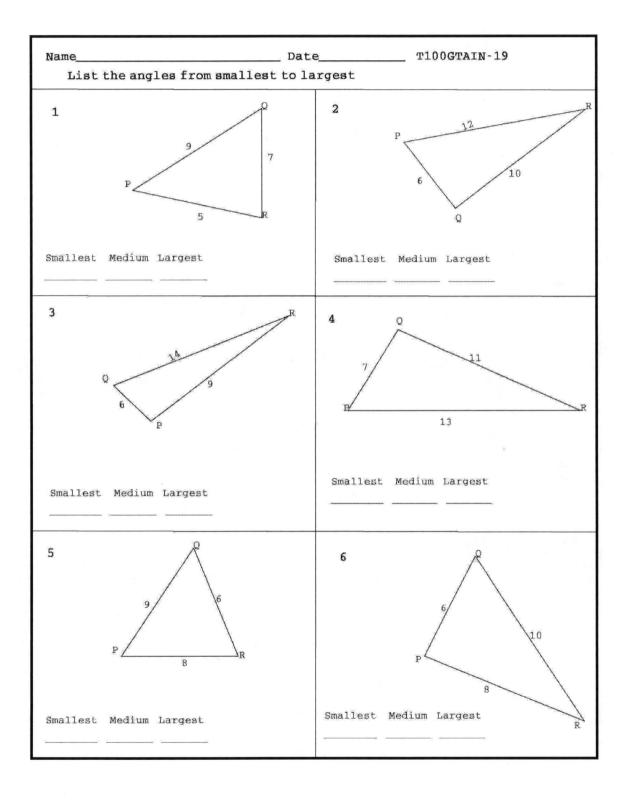

1

Q

9

7

P

5

R

Smallest Medium Largest

_____ _____ _____

2

R

P

12

6

10

Q

Smallest Medium Largest

_____ _____ _____

3

R

14

Q

9

6

P

Smallest Medium Largest

_____ _____ _____

4

Q

7

11

P

13

R

Smallest Medium Largest

_____ _____ _____

5

Q

9

6

P

8

R

Smallest Medium Largest

_____ _____ _____

6

Q

6

10

P

8

R

Smallest Medium Largest

_____ _____ _____

Name_____ Date_____ T80GTA-11

Find the unknown values.

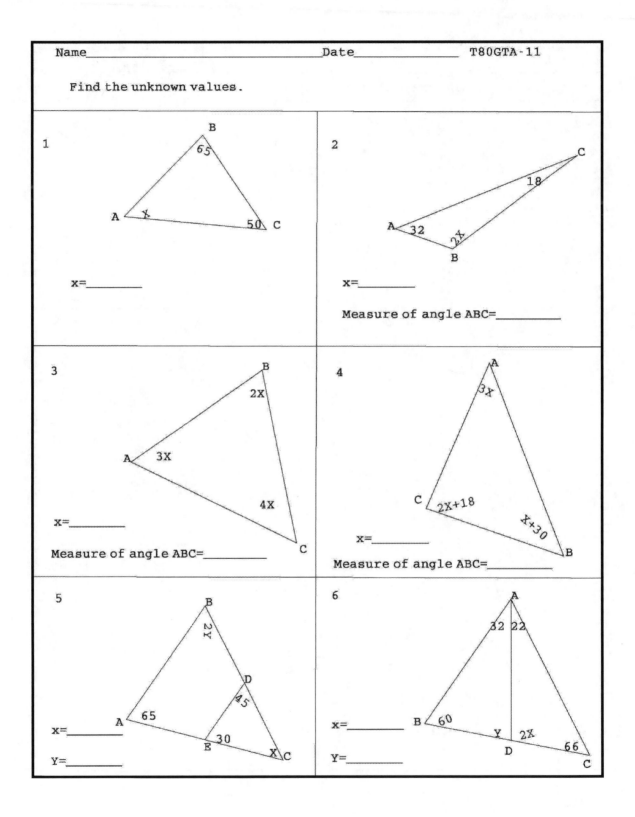

1

x=_____

2

x=_____

Measure of angle ABC=_____

3

x=_____

Measure of angle ABC=_____

4

x=_____

Measure of angle ABC=_____

5

x=_____

Y=_____

6

x=_____

Y=_____

Name_____ Date_____ T80GTEA-12

Find the unknown value

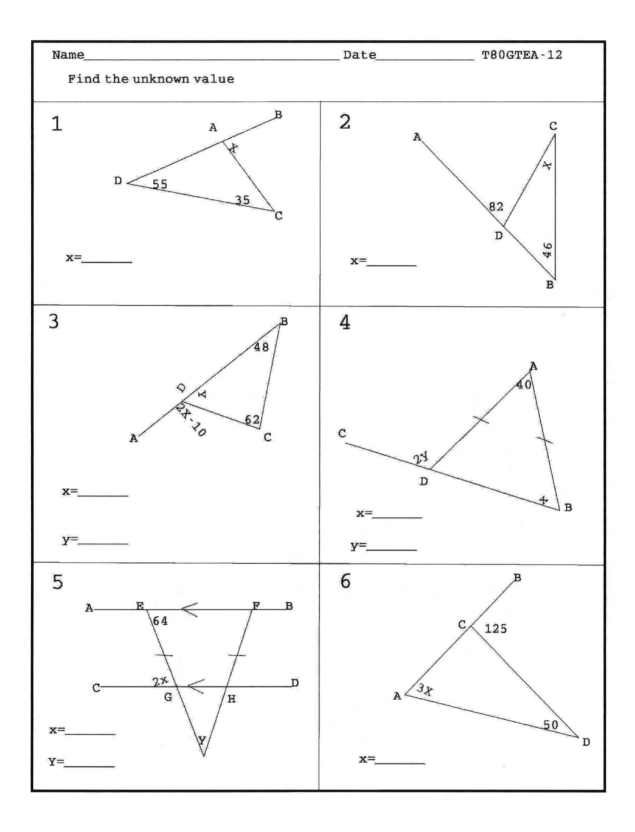

1

B
A
x
D   55
35
C

x=_____

2

A
C
x
82
D
46
B

x=_____

3

B
48
D Y
2X-10
A
62
C

x=_____

y=_____

4

A
40
C
2Y
D
x
B

x=_____

y=_____

5

A   E   F   B
64
C   2x   D
G   H
V

x=_____

Y=_____

6

B
C   125
A   3X
50
D

x=_____

Name_____ Date_____ T80PL-10

For the problems below, find the unknown value.

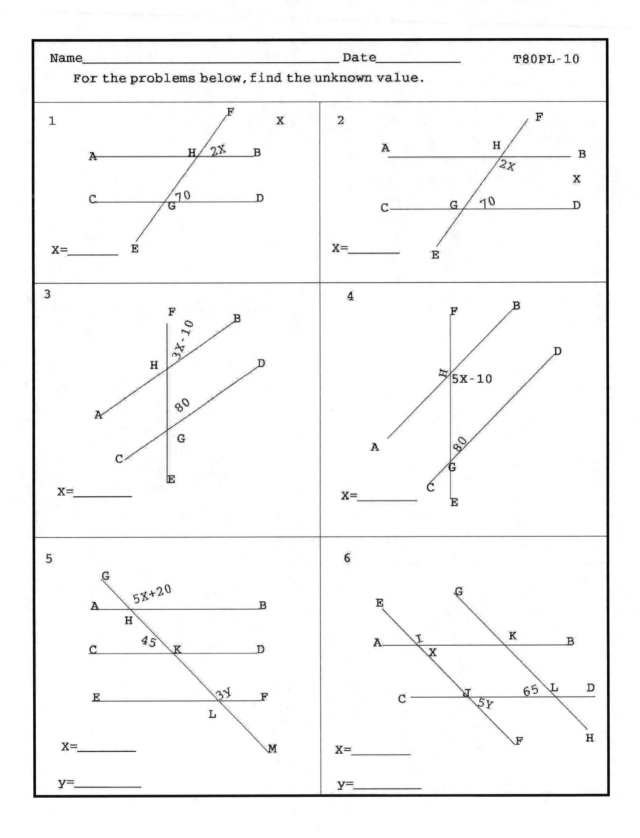

1

A ———— H 2X —— B

C ——— G 70 ——— D

F
X
E

X=_____

2

A ———— H ———— B
2X
C ——— G 70 ——— D
X
F
E

X=_____

3

F
B
H 3X-10
A
80
C G
E

X=_____

4

B
F
H 5X-10
A
D
80
C G
E

X=_____

5

G
A 5X+20 B
H
C 45 K D
E 3Y F
L
M

X=_____

Y=_____

6

E
G
A I K B
X
C J 65 L D
5Y
F
H

X=_____

Y=_____

Name_____ Date_____

Find the unknown value

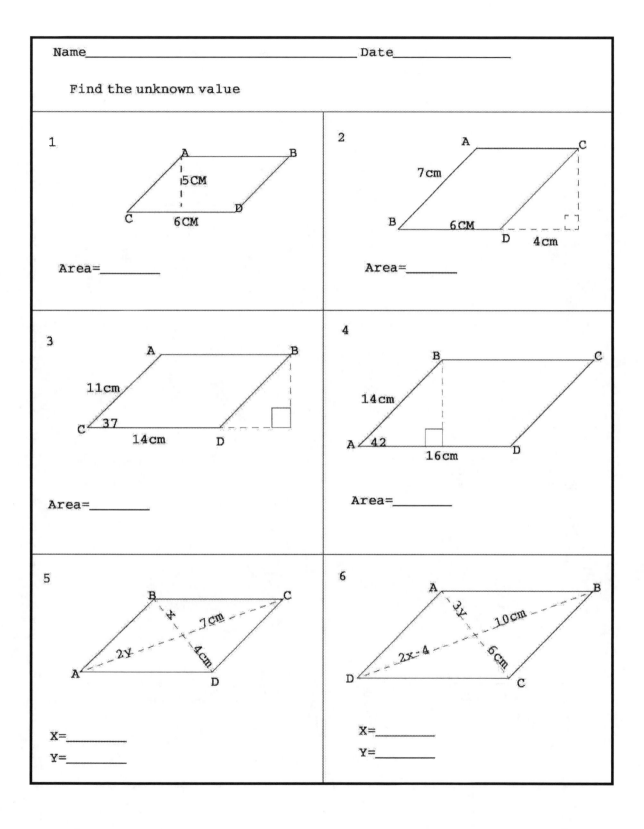

1

A   B
5CM
C   6CM   D

Area=_____

2

A   C
7cm
B   6CM   D   4cm

Area=_____

3

A   B
11cm
C 37   14cm   D

Area=_____

4

B   C
14cm
A 42   16cm   D

Area=_____

5

B   C
x   7cm
A   2y   4cm   D

X=_____
Y=_____

6

A   B
3y   10cm
D   2x-4   6cm   C

X=_____
Y=_____

Name_____ Date_____

For the trapizoids (Trapiziums) below, find the unknown

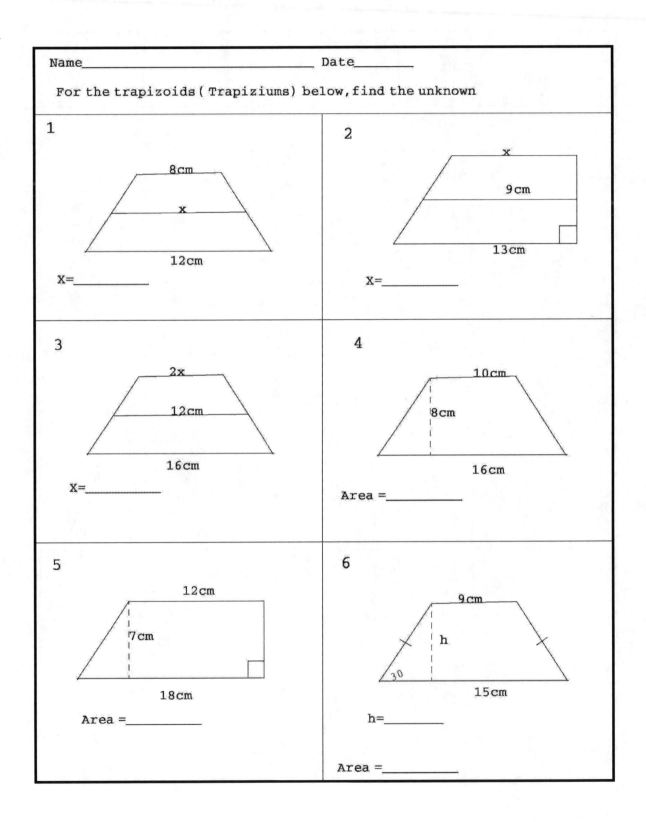

1
8cm
x
12cm
X=_____

2
x
9cm
13cm
X=_____

3
2x
12cm
16cm
X=_____

4
10cm
8cm
16cm
Area =_____

5
12cm
7cm
18cm
Area =_____

6
9cm
h
30
15cm
h=_____

Area =_____

Name_____ Date_____

Find the ratios of the indicated angles.

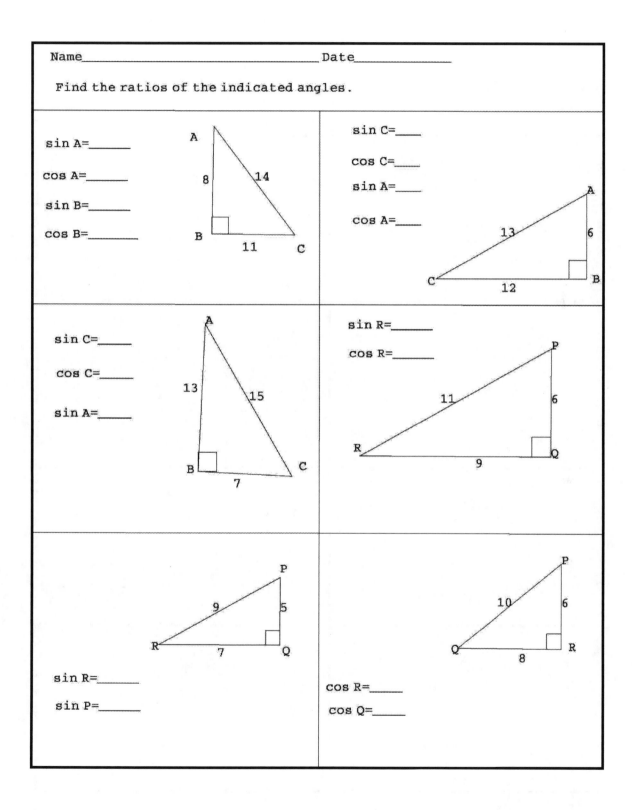

sin A=_____

cos A=_____

sin B=_____

cos B=_____

A

8   14

B ⌐ 11 C

---

sin C=____

cos C=____

sin A=____

cos A=____

A

13   6

C ⌐ 12 B

---

sin C=_____

cos C=_____

sin A=____

A

13   15

B ⌐ 7 C

---

sin R=_____

cos R=_____

P

11   6

R ⌐ 9 Q

P

9   5

R 7 Q ⌐

sin R=_____

sin P=_____

P

10   6

Q 8 R ⌐

cos R=_____

cos Q=_____

Name_____ Date _____

Use table or calculator to find the angle measure of the given ratios. Round answers to the nearest degree.

1. $\sin^{-1}(0.2356) =$ _____

2. $\sin^{-1}(0.6784) =$ _____

3. $\cos^{-1}(0.6780) =$ _____

4. $\cos^{-1}(0.2654) =$ ____

5. $\sin^{-1}(0.5) =$ ___

6. $\cos^{-1}(0.5) =$ _____

7. $\tan^{-1}(2.035) =$ _____

8. $\tan^{-1}(3.0564) =$ _____

Name_____ Date_____

Find the unknown value

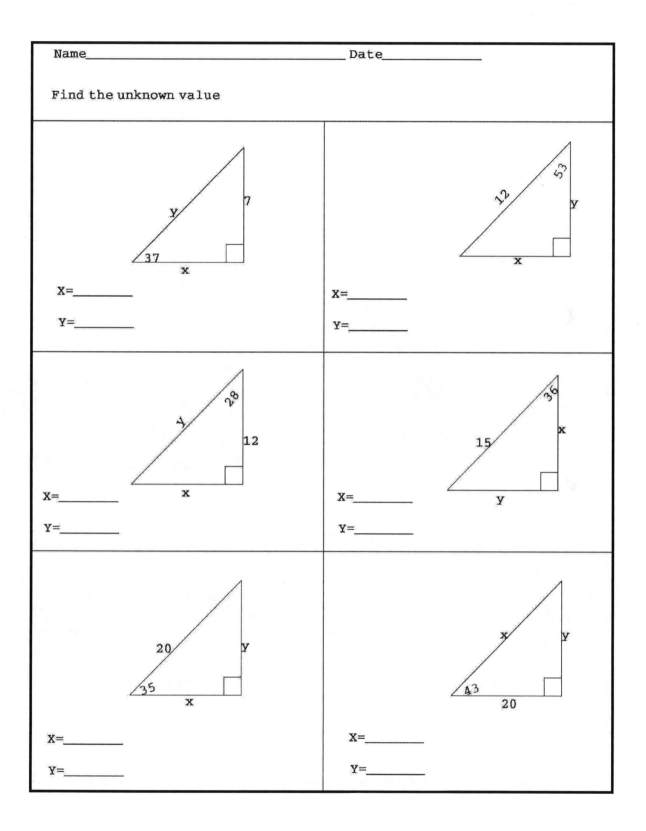

X=_____

Y=_____

X=_____

Y=_____

X=_____

Y=_____

X=_____

Y=_____

X=_____

Y=_____

X=_____

Y=_____

Name_____ Date_____

Find the unknown value

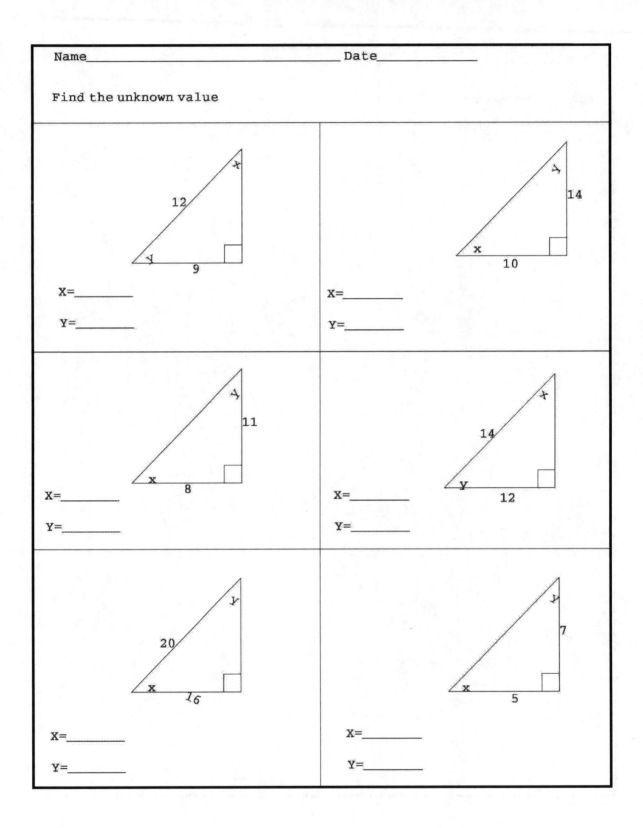

X=_____

Y=_____

X=_____

Y=_____

X=_____

Y=_____

X=_____

Y=_____

X=_____

Y=_____

X=_____

Y=_____

Name_____ Date _____

Convert the given angle to radian measure.

1. $\theta = 120°$	2. $\theta = 60°$
3. $\theta = 150°$	4. $\theta = 240°$
5. $\theta = 300°$	6. $\theta = -225°$
7. $\theta = -210°$	**8.** $\theta = 470°$

Name_____ **Date** _____

Convert each angle in radians to degrees.

1. $\dfrac{\pi}{4}$ *radian*	2. $\dfrac{3\pi}{4}$ *radian*
3. π *radian*	4. $\dfrac{3\pi}{2}$ *radians*
5. $\dfrac{7\pi}{6}$ *radians*	6. $\theta = -225°$ $\dfrac{7\pi}{4}$ *radians*
7. $-\dfrac{\pi}{6}$ *radian*	**8.** $-\dfrac{\pi}{3}$ *radian*

Name_____ Date_____

Find the angle and the arc measures

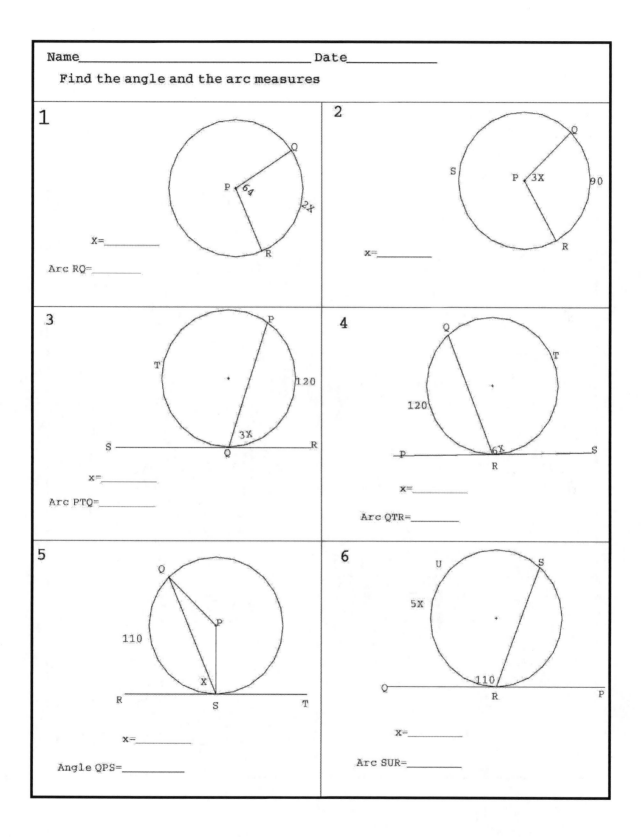

1

X=_____

Arc RQ=_____

2

x=_____

3

x=_____

Arc PTQ=_____

4

x=_____

Arc QTR=_____

5

x=_____

Angle QPS=_____

6

x=_____

Arc SUR=_____

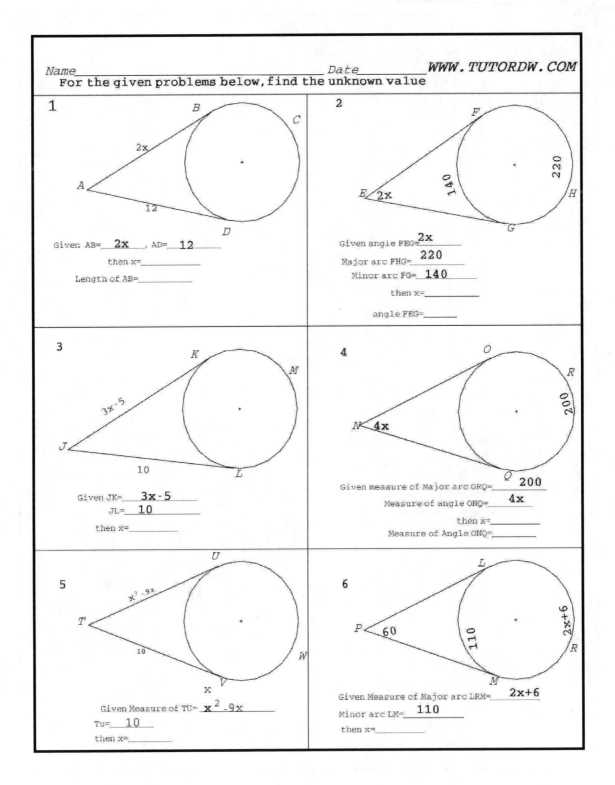

For the given problems below, find the unknown value

1 Given AB=___2x___, AD=___12___
then x=_____
Length of AB=_____

2 Given angle FEG=___2x___
Major arc FHG=___220___
Minor arc FG=___140___
then x=_____
angle FEG=_____

3 Given JK=___3x-5___
JL=___10___
then x=_____

4 Given measure of Major arc ORQ=___200___
Measure of angle ONQ=___4x___
then x=_____
Measure of Angle ONQ=_____

5 Given Measure of TU=___x^2-9x___
TU=___10___
then x=_____

6 Given Measure of Major arc LRM=___2x+6___
Minor arc LM=___110___
then x=_____

Name_____Date_____   WWW.tutordw.com

For the given problems below, find the unknown value

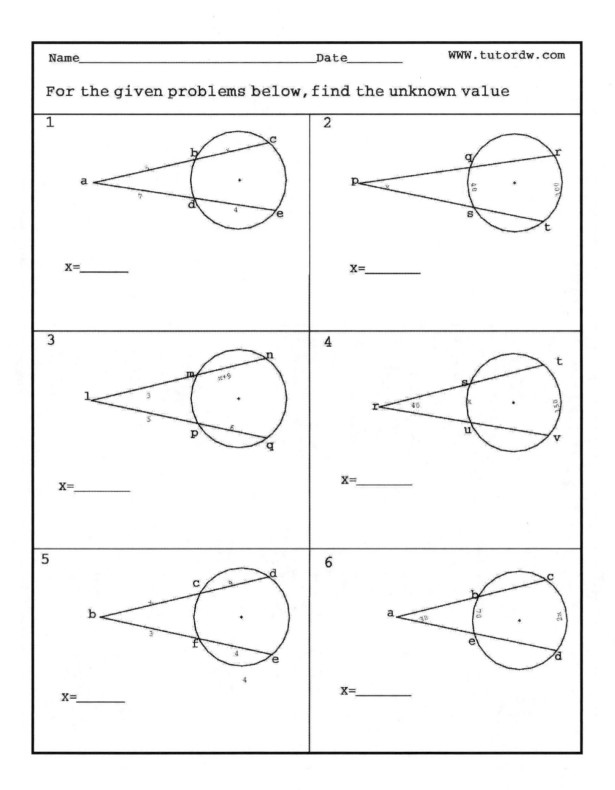

1

X=_____

2

X=_____

3

X=_____

4

X=_____

5

X=_____

6

X=_____

For the problems given below, find the unknown value

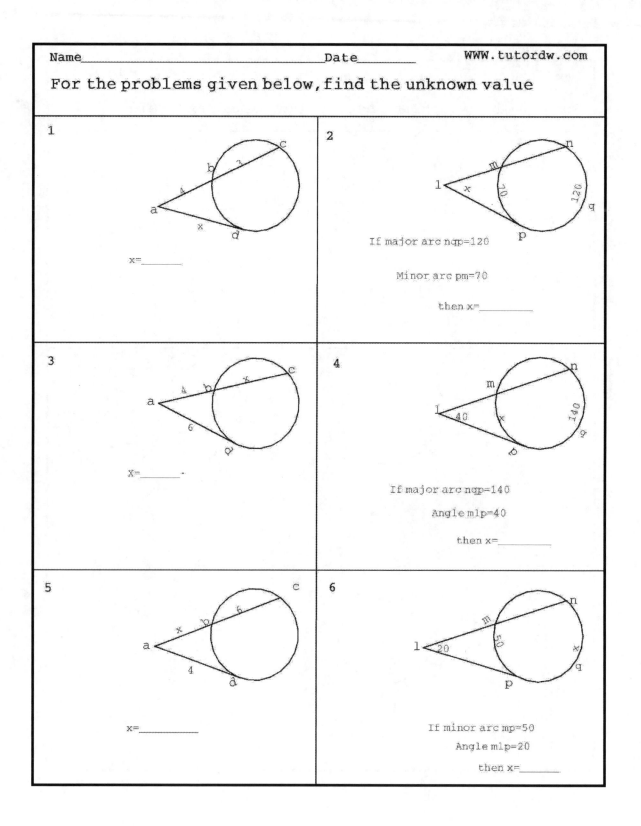

1

x=_____

2

If major arc nqp=120

Minor arc pm=70

then x=_____

3

X=_____ -

4

If major arc nqp=140

Angle mlp=40

then x=_____

5

x=_____

6

If minor arc mp=50

Angle mlp=20

then x=_____

Name_____ Date_____ T100GC-16

Find the unknown value for the problems below.

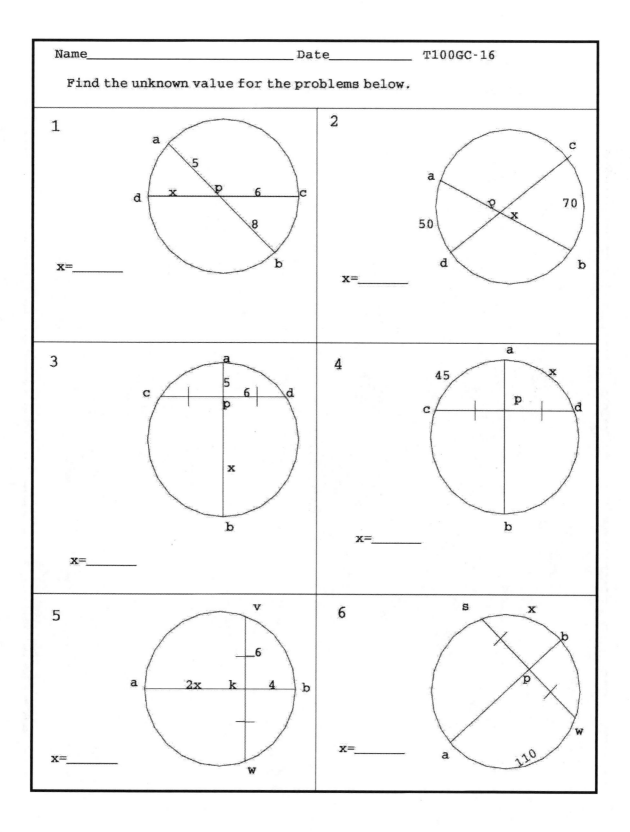

1

5

x   p   6

d   x   c

8

b

x=_____

2

c

a

o

50   x   70

d   b

x=_____

3

a

5

6

c   p   d

x

b

x=_____

4

a

45   x

p

c   d

b

x=_____

5

v

6

a   2x   k   4   b

w

x=_____

6

s   x

b

p

a   110   w

x=_____

Name_____ Date_____ T100GC-20

For the problems below, find the unknown.

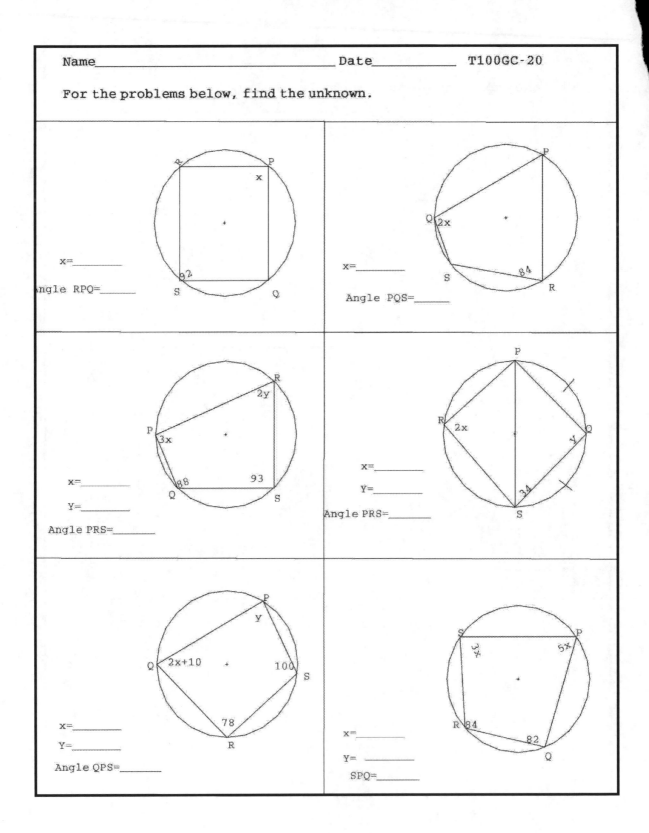

x=_____

Angle RPQ=_____

92

x=_____

Angle PQS=_____

84

x=_____

Y=_____

Angle PRS=_____

2y 3x 88 93

x=_____

Y=_____

Angle PRS=_____

2x 34

x=_____

Y=_____

Angle QPS=_____

2x+10 100 78 y

x=_____

Y= _____

SPQ=_____

3x 5x 84 82

Name_____ Date_____ T100GCT-   14

For the given problems below, find the unknown value

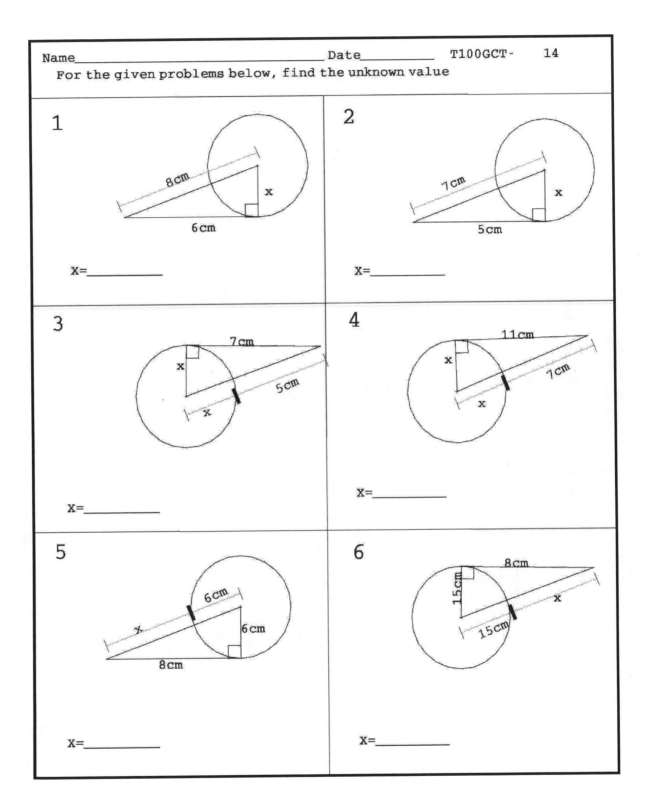

1

8cm

x

6cm

X=_____

2

7cm

x

5cm

X=_____

3

7cm

x

5cm

x

X=_____

4

11cm

x

7cm

x

X=_____

5

6cm

x

6cm

8cm

X=_____

6

8cm

15cm

x

15cm

X=_____